The Regulation of Embryo Testing in Australia

Michelle de Souza

The Regulation of Embryo Testing in Australia

A Principles-based Approach

Springer

Michelle de Souza
Law School
University of Western Australia
Crawley, WA, Australia

ISBN 978-981-96-5901-2 ISBN 978-981-96-5902-9 (eBook)
https://doi.org/10.1007/978-981-96-5902-9

This Springer imprint is published by the registered company Springer Nature Singapore Pte Ltd.
The registered company address is: 152 Beach Road, #21-01/04 Gateway East, Singapore 189721, Singapore

Contents

Chapter 1
Embryo Testing: An Introduction

Abstract This introductory chapter outlines the rationale for embryo testing and the development of this technology. It also discusses the importance of regulation in this area, and provides an overview of the chapters which follow.

1.1 Introduction

The birth of Louise Brown in July 1978 heralded the start of what may be termed the 'reproductive revolution'.[1] While for couples hindered by infertility the world's first successful birth through in vitro fertilisation (IVF) meant new hope, for others it conjured images of monsters being grown in test tubes. IVF is the most common fertility treatment and involves the creation of an embryo outside the human body. It is also the backbone of the assisted reproductive technology (ART) industry. The first step of an IVF procedure involves the prescription of ovarian stimulation drugs. These drugs enable the prospective mother's ovaries to produce additional eggs, as naturally the ovaries will only produce one per month. The next step in the procedure is egg retrieval. Eggs are retrieved from the ovaries by the insertion of a thin needle into a woman's vagina.[2] Once the eggs are retrieved, they are mixed with either fresh or cryopreserved (frozen) sperm.[3] At this stage it is hoped that at least some of the eggs will be fertilised and become embryos. It is at this point in the IVF cycle that embryo testing, if required, may be undertaken, and it is the regulation of this testing that will be the subject of this book. If embryo testing is not required, one or possibly two embryos will be implanted into the woman.

In this book, embryo testing is used as an umbrella term to describe the diagnostic tests carried out on human embryos for the purposes of selecting which embryo, or embryos, will be implanted into an individual. Four kinds of embryo

[1] Lee R, Morgan D (2001) Human Fertilisation and Embryology: Regulating the Reproductive Revolution. Blackstone Press, London. See also Brazier M (1998) Regulating the Reproduction Business? Medical Law Review 7: 166, 167.

[2] Deech R, Smajdor A, (2007) From IVF to Immortality. Oxford University Press, 18.

[3] Ibid 19.

M. de Souza, *The Regulation of Embryo Testing in Australia*,
https://doi.org/10.1007/978-981-96-5902-9_1

testing are currently permitted in Australia: preimplantation genetic testing for monogenic conditions (PGT-M),[4] preimplantation genetic testing for aneuploidy (PGT-A),[5] preimplantation genetic testing for structural relocations (PGT-SR)[6] and human leukocyte antigen typing (HLA typing).[7] PGT-M[8] involves testing an embryo for a medical condition[9] and allows for the deselection of embryos that have that condition. The purpose for which PGT-A is undertaken is slightly different to that of PGT-M. Rather than testing for a specific condition, as in PGT-M, PGT-A involves screening an embryonic cell in order to identify the number of chromosomes.[10] As embryos with an abnormal chromosome number are likely to become a future child with a disability, for example Down Syndrome,[11] or lead to a miscarriage,[12] such embryos can be discarded prior to implantation. PGT-M and PGT-A may also be used to identify the sex of an embryo, thereby enabling the avoidance of sex-linked conditions. As both PGT-M and PGT-A are used for the avoidance of disability, references to PGT-M in the following chapters should be taken to include PGT-A.

PGT-SR detects structural rearrangements in an embryo's chromosomes, sometimes referred to as translocations.[13] These translocations may be balanced, meaning that no genetic material is gained or lost, or unbalanced, resulting in more or less genetic material. Therefore, unlike PGT-A which identifies the number of chromosomes, PGT-SR focuses on the structure of the chromosomes. PGT-SR is performed where prospective parents are the carriers of translocations, to avoid the implantation of embryos with unbalanced chromosomal rearrangements. As with embryos that have an unbalanced chromosome number, embryos with unbalanced chromosomal rearrangements may develop into a child with a disability or lead to a miscarriage. The other kind of embryo testing, HLA typing, involves testing an embryo for tissue

[4] National Health and Medical Research Council, *Ethical Guidelines on the Use of Assisted Reproductive Technology in Clinical Practice and Research 2017* (2017) 73.

[5] Ibid.

[6] Ibid.

[7] Ibid 52.

[8] Until 2017, PGT-M was referred to as PGD (preimplantation genetic diagnosis). Readers will notice the term PGD in references to pre-2017 literature. See generally Zegers-Hochschild F et al. (2017) The International Glossary on Infertility and Fertility Care, 2017. Fertility and Sterility 108(3): 393.

[9] Sullivan-Pyke C, Dokras A (2018) Preimplantation Genetic Screening and Preimplantation Genetic Diagnosis. Obstetrics and Gynecology Clinics of North America 45: 113, 115.

[10] Ibid.

[11] Down Syndrome is caused by the presence of three number 21 chromosomes, known as trisomy 21. See generally, Pope BD, Gilbert DM (2014) Up and Down in Down's Syndrome. Nature 508: 323, 323.

[12] For example, trisomies 16 and 22.

[13] Campos-Galindo I, Peinado V (2022) Preimplantation Genetic Testing for Structural Rearrangements. In: Simón C, Rubio C (eds), Handbook of Genetic Diagnostic Technologies in Reproductive Medicine: Improving Patient Success Rates and Infant Health. Taylor & Francis 136, 141.

type. HLA typing allows for the positive selection of an embryo that is a tissue match for a sick family member.[14]

In addition to these sanctioned uses of embryo testing, there are prohibited uses such as non-medical sex selection and testing with a view to choosing an embryo that will become a child with a disability. In the not too distant future, the possibilities offered by embryo testing will expand beyond what is currently possible. These unknown possibilities bring with them a wealth of regulatory challenges. At present, the current regulatory framework is not equipped to deal with these challenges, or embrace the opportunities presented by new technological developments in embryo testing. Accordingly, the key questions addressed by this book are what limits should be placed on prospective parents' use of embryo testing and what form should the regulatory framework take?

Embryo testing, like IVF, involves the creation of multiple embryos, not all of which will be implanted into a woman. It therefore involves the destruction of embryos. As some religions believe life begins at conception,[15] the creation of embryos that will not have the opportunity to become a child is objectionable. This book begins from the assumption that it is acceptable to create embryos knowing that they may be destroyed.[16] Therefore, this issue will not be dealt with in this book.

This chapter aims to provide an overview of and background to embryo testing. It begins by considering why PGT and HLA typing are necessary and how they were developed. The direction these technologies may take in the future will then be discussed. As the focus of this book is the regulation of embryo testing, the question of why embryo testing should be regulated will be addressed. Lastly, this chapter provides an overview of each subsequent chapter of the book.

1.2 The Need for PGT

Prior to the advent of PGT, prospective parents living with an inherited medical condition were faced with three options: they could decide the risk of conceiving a child with a genetic condition was too great and choose not to have genetically

---

[14] See generally Brown L, Gaspar HB, 'Preimplantation Genetic Diagnosis and HLA-Typing: The Potential for Selection of Unaffected HLA Matched Siblings' in El-Toukhy T, Braude P (eds), *Preimplantation Genetic Diagnosis in Clinical Practice* (Springer, 2014) 141.

[15] The religious group most often associated with this belief is the Catholic Church. See eg, Pope John Paul II, *Evangelium Vitae* (25 March 1995) The Holy See < http://w2.vatican.va/content/john-paul-ii/en/encyclicals/documents/hf_jp-ii_enc_25031995_evangelium-vitae.html >. However, the belief that life begins at fertilisation is also held by other faiths such as Buddhists, Hindus and Sikhs. See eg, Keown D (2009) Buddhist Bioethics. In: Singer PA, Viens AM (eds) The Cambridge Textbook of Bioethics. Cambridge University Press, 391, 393; Coward H and Sidhu T (2009) Hindu and Sikh Bioethics. In: Singer PA and Viens AM (eds) The Cambridge Textbook of Bioethics. Cambridge University Press, 403, 406.

[16] For a brief discussion of why it may be acceptable to create embryos that may be discarded, see Pennings G, de Wert G (2012) Preimplantation Genetic Diagnosis. In: Chadwick R (ed) Encyclopedia of Applied Ethics vol 2, 576, 577.

related children. Alternatively, they could take the risk that their child will inherit the disorder. Should the prospective parents decide to accept the risk, pre-natal diagnosis was an option for those willing to consider aborting an affected foetus. Lastly, they could use a sperm donor who was neither a carrier nor a sufferer of the disorder. Embryo testing provides a way for parents who were previously faced with these choices to have a genetically related child without the attendant risks.

PGT-M prevents the birth of children with genetic disorders. It is used to test for and select against single gene disorders such as cystic fibrosis,[17] Tay-Sachs[18] and Fragile X syndrome.[19] Single gene disorders are conditions caused by the mutation of a single gene in a person's chromosomes. Such disorders may be autosomal dominant conditions or autosomal recessive conditions. If a future child is at risk of inheriting an autosomal dominant condition, only one copy of the faulty gene is required for the child to inherit the disorder.[20] In the case of autosomal dominant conditions, such as Huntington's disease,[21] assuming only one parent carries the faulty gene, there is a 50% chance a child will inherit the disorder.[22] Should both parents have the disorder, assuming each only carries one copy of the mutation, there is a 75% chance their child will inherit the condition. On the other hand, a child will only suffer from an autosomal recessive disorder, such as cystic fibrosis, if two copies of the faulty

---

[17] Cystic fibrosis is caused by a mutation in a gene that produces a protein called CFTR. One of the functions of CFTR is to clear the airways of mucus. Thus, a failure to produce this protein can result in in a build-up of mucus in the lungs, airways and digestive system. This in turn causes breathing difficulties and lung infections. The mucus also affects the pancreas, causing digestive problems. As there is no cure for cystic fibrosis, the condition is managed through a combination of medication and physiotherapy. See generally O'Sullivan BP, Freedman SD (2009) Cystic Fibrosis. The Lancet 373: 1891; Stuart Elborn J (2016) Cystic Fibrosis. Lancet 388: 2519.

[18] Tay-Sachs disease inhibits the production of an enzyme known as Hex-A. Without this enzyme, a lipid (GM2 ganglioside) builds up in the cells of the sufferer. This lipid is particularly toxic to the nerve cells of the brain and causes neurological degeneration. Death may occur in infancy or early childhood (usually three to four years old). Crawford D (2018) Tay-Sachs Disease. Nursing Children and Young People 30(5): 17, 17. See also Altarescu G et al. (2007) Simultaneous Preimplantation Genetic Diagnosis for Tay–Sachs and Gaucher Disease. Reproductive BioMedicine Online 15: 83.

[19] It has been suggested that Fragile X syndrome is the most common cause of inherited intellectual disability in males. While it is possible for females to inherit the condition, they are usually less severely affected than males. Garber KB, Visootsak J, Warren ST (2008) Fragile X Syndrome. European Journal of Human Genetics 16: 666, 666–667.

[20] Cook J (2013) Genes in Families. In: Rimoin D, Pyeritz R, Korf B (eds) Emery and Rimoin's Principles and Practice of Medical Genetics. Academic Press, 1, 5.

[21] Huntington's disease is a progressive neurodegenerative disorder. Individuals with the condition experience motor difficulties and a decline in their cognitive functions, resulting in psychiatric disorders, dementia and depression. Unlike the other single gene disorders discussed in this chapter, Huntington's disease is a late onset condition; this means that sufferers do not experience symptoms until they are in their mid-thirties to mid-forties. Moutou C, Gardes N, Viville S (2004) New Tools for Preimplantation Genetic Diagnosis of Huntington's Disease and their Clinical Applications. European Journal of Human Genetics 12: 1007, 1007; Van Rij MC et al. (2012) Preimplantation Genetic Diagnosis (PGD) for Huntington's Disease: The Experience of Three European Centres. European Journal of Human Genetics 20: 368, 368.

[22] Flinter F, Stewart F (2014) Basic Genetics for PGD. In: El-Toukhy T, Braude P (eds) Preimplantation Genetic Diagnosis in Clinical Practice. Springer, 9, 18.

gene are passed onto the child.[23] Therefore, if each parent carries one copy of the mutation, there is a one in four chance their child will suffer from the condition.

Some single gene disorders are also sex-linked disorders. This means the faulty gene occurs on one of the sex chromosomes. As females have two X chromosomes, they are not affected by such disorders, provided there is a normal copy of the gene on the other X chromosome.[24] For example, haemophilia[25] is often referred to as a sex-linked condition. Therefore, in a couple where the mother is a carrier of the haemophilia mutation but the father is not a haemophilia sufferer, there is a 25% chance that any sons the couple have will be affected by the condition.[26] This is because a son will inherit the haemophilia gene on his X chromosome from his mother and does not have a second X chromosome containing a normal copy of the gene. While there is also a 25% chance that a daughter will be a carrier of the condition, she will not be a sufferer. As a result, when prospective parents are at risk of passing on a sex-linked condition, rather than testing for a specific disorder, PGT is used to identify the sex of the embryos.

A person with a single gene disorder may require significant medical treatment throughout their life. Some disorders may also cause death in early childhood[27] or may be incurable.[28] PGT enables parents to avoid passing on such genetic disorders to their offspring. Embryo testing also confers an additional benefit on these families, as future generations will not need to worry about their offspring inheriting the medical condition.

1.3 The Need for HLA Typing

Human leukocyte antigen typing (HLA typing), or tissue typing, allows the parents of a sick child to have a second child who can potentially provide a cure for the first child's illness by donating stem cells.[29] Stem cells are transplanted from the umbilical cord of the newborn and thus the newborn does not experience any pain as

---

[23] Cook J (2013) Genes in Families. In: Rimoin D, Pyeritz R, Korf B (eds) Emery and Rimoin's Principles and Practice of Medical Genetics. Academic Press, 1, 7.

[24] Ibid 10.

[25] Haemophilia is a disorder caused by a low level of one of the essential clotting factors. As the blood is not able to clot properly, individuals with the condition experience internal bleeding as well as excessive bleeding following a cut. Haemophilia Foundation Australia, Haemophilia (August 2021) < https://www.haemophilia.org.au/about-bleeding-disorders/haemophilia >.

[26] Cook J (2013) Genes in Families. In: Rimoin D, Pyeritz R, Korf B (eds) Emery and Rimoin's Principles and Practice of Medical Genetics. Academic Press, 1, 10.

[27] Tay-Sachs disease and Niemann-Pick disease type A are examples of conditions where death occurs in early childhood.

[28] Examples of incurable single gene disorders include cystic fibrosis and fragile X syndrome.

[29] See generally Devolder K (2005) Preimplantation HLA Typing: having children to save our loved ones. Journal of Medical Ethics 31: 582, 582.

a consequence of the transplant.[30] Children who have been selected for their tissue type are often referred to as 'saviour siblings'.[31]

A stem cell transplant following the birth of an HLA matched child may provide a cure for disorders of the haematopoietic system.[32] Such disorders include various types of anaemia,[33] thalassaemia and leukaemia. In some cases, a haematopoietic stem cell transplant is the only option for curing a haematopoietic system disorder.[34]

Each person has two different HLA haploid genotypes,[35] with one group of human leukocyte antigens inherited from each parent. Therefore, when parents decide to have a second child, there is a one in four chance that the child will be an HLA match for their sick sibling.[36] Tissue typing plays an important role in the creation of a stem cell donor, because a sick child has the best chance of survival if they have a stem cell donor who is an exact HLA match.[37] While it is possible to find an unrelated stem cell donor, minor differences in HLA type have a detrimental impact upon the success of the transplant.[38]

HLA typing may be used in conjunction with PGT-M to select an embryo that is not only unaffected by a genetic condition, but also a tissue match for an existing child.[39] Where PGT-M is required in conjunction with HLA typing, the chances of creating an embryo that is both unaffected by the genetic disorder and an HLA match are slim. For example, in the case of autosomal recessive disorders, if both parents are carriers there is a 75% chance that any child born will be unaffected (although they may be a carrier). This means that if parents are simply looking to avoid passing on an

---

[30] Brown L, Gaspar HB (2014) Preimplantation Genetic Diagnosis and HLA-Typing: The Potential for Selection of Unaffected HLA Matched Siblings. In: El-Toukhy T, Braude P (eds) Preimplantation Genetic Diagnosis in Clinical Practice. Springer, 141, 149.

[31] See, eg, ibid 148; Devolder K (2005) Preimplantation HLA Typing: having children to save our loved ones. Journal of Medical Ethics 31: 582, 583; Liu CK (2007) 'Saviour Siblings'? The Distinction between PGD with HLA Tissue Typing and Preimplantation HLA Tissue Typing. Journal of Bioethical Inquiry 4: 65; Spriggs M, Savulescu J (2002) "Saviour siblings". Journal of Medical Ethics 28: 289.

[32] Brown L, Gaspar HB (2014) Preimplantation Genetic Diagnosis and HLA-Typing: The Potential for Selection of Unaffected HLA Matched Siblings. In: El-Toukhy T, Braude P (eds) Preimplantation Genetic Diagnosis in Clinical Practice. Springer, 141, 141.

[33] For example, Fanconi anaemia.

[34] Brown L, Gaspar HB (2014) Preimplantation Genetic Diagnosis and HLA-Typing: The Potential for Selection of Unaffected HLA Matched Siblings. In: El-Toukhy T, Braude P (eds) Preimplantation Genetic Diagnosis in Clinical Practice. Springer, 141, 142.

[35] A haploid genotype is a set of genes found on a single chromosome.

[36] Brown L, Gaspar HB (2014) Preimplantation Genetic Diagnosis and HLA-Typing: The Potential for Selection of Unaffected HLA Matched Siblings. In: El-Toukhy T, Braude P (eds) Preimplantation Genetic Diagnosis in Clinical Practice. Springer, 141, 145. As there is a 75% chance that a child conceived naturally will not be a tissue match, HLA typing saves parents from making the difficult decision of whether to terminate the pregnancy and try again.

[37] Ibid.

[38] Ibid 146.

[39] See, eg, Bielorai B et al. (2004) Successful Umbilical Cord blood transplantation for Fanconi Anemia Using Preimplantation Genetic Diagnosis for HLA-matched Donor. American Journal of Hematology 77: 397.

autosomal recessive condition to their offspring there is a 75% chance of a successful outcome from PGT-M. However, when used in conjunction with HLA typing, where there is only a one in four chance of an HLA matched embryo, the likelihood of a successful outcome decreases to three in sixteen, or 18.75%.[40] Further, it should be noted that these success rates refer to the likelihood of finding a suitable embryo and thus the chance of a live birth is even lower.

HLA typing can also be carried out in the absence of PGT-M for conditions that are not inherited. An example of such a condition is Diamond-blackfan anaemia. While Diamond-blackfan anaemia may be inherited, in more than 50% of cases sufferers do not have a familial history of the disorder. The use of HLA typing in the absence of PGT has been controversial[41] and the ethics of subjecting an embryo to tests that are solely designed to help another person are examined in the following chapters.

1.4 The Development of PGT

PGT was originally referred to as preimplantation diagnosis and given the acronym PID. However, the acronym PID was already being used for pelvic inflammatory disease. Thus to avoid confusion it became known as preimplantation genetic diagnosis or PGD.[42] Then, in 2017, the World Health Organization renamed it PGT.[43] This new nomenclature also included renaming preimplantation genetic screening; now known as PGT-A.

Initially developed for use in animals, PGT was first used in 1968 for the determination of sex following the biopsy of five-day-old rabbit embryos.[44] However, the development of PGT for clinical application in humans stalled, in part due to a lack of human embryos for experimentation.[45] Then, in 1978, Louise Brown became the first baby to be born following IVF. IVF not only gave hope to the infertile, but also

---

[40] Brown L, Gaspar HB (2014) Preimplantation Genetic Diagnosis and HLA-Typing: The Potential for Selection of Unaffected HLA Matched Siblings. In: El-Toukhy T, Braude P (eds) Preimplantation Genetic Diagnosis in Clinical Practice. Springer, 141, 147.

[41] See, eg, Human Fertilisation and Embryology Authority, 'HFEA confirms that HLA tissue typing may only take place when PGD is required to avoid a serious genetic disorder' (Press Release, 1 August 2002) < http://www.hfea.gov.uk/935.html > ; Sheldon S, Wilkinson S (2004) Hashmi and Whitaker: An Unjustifiable and Misguided Distinction? Medical Law Review 12(2): 137.

[42] Harper J (2009) Introduction to Preimplantation Genetic Diagnosis. In: Harper J (ed) Preimplantation Genetic Diagnosis. Cambridge, 2nd ed, 1, 5.

[43] Harper J C, SenGupta S (2019) Preimplantation Genetic Testing. In: Stewart J A (ed) Subfertility, Reproductive Endocrinology and Assisted Reproduction. Cambridge Medicine. Cambridge University Press, 206–213, 206.

[44] Gardner RL, Edwards RG (1968) Control of the Sex Ratio at Full Term in the Rabbit by Transferring Sexed Blastocysts. Nature 218: 346.

[45] Theodosiou AA, Johnson MH (2011) The Politics of Human Embryo Research and the Motivation to Achieve PGD. Reproductive Biomedicine Online 22: 457, 459.

provided a new source of human embryos for experimentation. Despite this, it still took 12 years for the first successful human pregnancy following PGT.[46]

PGT involves two stages. First, a cell is extracted from an early stage embryo, then that cell is subjected to genetic testing. To carry out diagnostic testing on a single cell requires a sensitive and sophisticated test and this is one of the reasons PGT proved so difficult to develop in humans.[47] In 1987, Marilyn Monk and Alan Handyside successfully carried out the first single cell diagnostic test. The test was carried out on a mouse that had been genetically engineered with a defect in its hypoxanthine phosphoribosyl transferase (HPRT) gene. In humans, the HPRT gene is associated with Lesch-Nyhan syndrome.[48] Monk and her team were thus able to test a single cell for the mouse model of Lesch-Nyhan syndrome.[49] This was the first time that analysing a single cell for a single gene defect was considered a feasible proposition.[50]

During the late 1980s, many teams around the world were trying to translate the successes of the animal experiments into clinical PGT that could be used for the diagnosis of human embryos. Scientists in London and New York worked on cleavage stage biopsy, which involved the extraction of cells from an embryo one to two days after fertilisation. In Chicago, Yury Verlinsky and his team were working on polar body biopsy.[51] The London team, led by Robert Winston and Alan Handyside, biopsied two to three day old embryos and found that the removal of cells from three day old embryos had little effect on the development of the embryo.[52] Using a polymerase chain reaction Handyside was able to identify male embryos, giving hope to couples at risk of transmitting a sex-linked disorder to their offspring.[53] A year later, Handyside and his team reported the first successful pregnancies following PGT.[54] These pregnancies were the result of the implantation of female embryos after PGT for the purposes of sex selection. In 1992, Handyside's team carried out the

[46] Handyside AH et al. (1990) Pregnancies from Biopsied Human Preimplantation Embryos Sexed by Y-Specific DNA Amplification. Nature 344(6268): 768.

[47] Harper J (2009) Introduction to Preimplantation Genetic Diagnosis. In: Harper J (ed) Preimplantation Genetic Diagnosis. Cambridge, 2nd ed, 1, 3.

[48] Lesch-Nyhan syndrome is a sex-linked autosomal recessive condition. Most individuals with the condition are unable to walk and experience behavioural problems. One of the characteristic symptoms of the disease is self-harm, where a sufferer bites their own hands, fingers, lips and cheeks. Nyhan WL (2014) Lesch-Nyhan Syndrome. In: Aminoff MJ, Daroff RB (eds) Encyclopedia of the Neurological Sciences. Academic Press, 2nd ed, 785, 785.

[49] Monk M et al. (1987) Preimplantation Diagnosis of Deficiency of Hypoxanthine Phosphoribosyl Transferase in a Mouse Model for Lesch-Nyhan Syndrome. The Lancet 330: 423.

[50] Harper J (2009) Introduction to Preimplantation Genetic Diagnosis. In: Harper J (ed) Preimplantation Genetic Diagnosis. Cambridge, 2nd ed, 1, 5.

[51] Ibid 5.

[52] Hardy K et al. (1990) Human Preimplantation Development In Vitro is Not Adversely Affected by Biopsy at the 8-cell Stage. Human Reproduction 5: 708.

[53] Handyside AH et al. (1989) Biopsy of Human Preimplantation Embryos and Sexing by DNA Amplification. The Lancet 333: 347.

[54] Handyside AH et al. (1990) Pregnancies from Biopsied Human Preimplantation Embryos Sexed by Y-Specific DNA Amplification. Nature 344(6268): 768.

first single cell diagnosis of cystic fibrosis.[55] Once again using a polymerase chain reaction for DNA amplification, Handyside was able to identify the cystic fibrosis mutation on single cells extracted from embryos. Of the three couples who took part in the trial, one became pregnant and gave birth to a baby girl who was neither a sufferer nor a carrier of cystic fibrosis.[56]

1.5 The Development of HLA Typing

The success of PGT led the way for the development of HLA typing. In 2001, Verlinsky and his team published their research on the use of HLA typing in conjunction with PGT to create a child who could provide a stem cell transplant for their older sibling who was suffering from Fanconi anaemia.[57] Fanconi anaemia is a rare, autosomal recessive, genetic disorder that affects a person's bone marrow. In many cases it leads to bone marrow failure, meaning that affected individuals are unable to produce any new blood cells. A stem cell transplant is the only way to cure Fanconi anaemia, as the donor's stem cells instigate the creation of blood cells and platelets again.[58]

Verlinsky's team performed PGT on cells from early embryos that had been created by a couple who both carried the mutation for Fanconi anaemia. The couple had a daughter who was a sufferer, and hoped that a stem cell transplant from a tissue matched sibling would cure her. As was commonplace by 2000, a polymerase chain reaction was carried out for the purposes of DNA amplification.[59] This enabled the identification of embryos that carried the Fanconi anaemia mutation and also allowed determination of HLA type. Four cycles of IVF generated 5 embryos that were neither affected by, nor carriers of Fanconi anaemia, as well as 19 embryos that were carriers. All 24 embryos were tested for HLA type and five of the carrier embryos proved to be matches. A successful pregnancy resulted and chorionic villus sampling confirmed that the embryo would not suffer from Fanconi anaemia and was an HLA match.[60] After the birth of the child, stem cells from the umbilical cord were transplanted to their older sibling, leading to 'haematopoietic reconstitution'.[61] The child, Adam Nash, became the world's first saviour sibling.[62]

[55] Handyside AH et al. (1992) Birth of a Normal Girl after In Vitro Fertilization and Preimplantation Diagnostic Testing for Cystic Fibrosis. New England Journal of Medicine 327: 905.

[56] Ibid 908.

[57] Verlinsky Y et al. (2001) Preimplantation Diagnosis for Fanconi Anemia Combined with HLA Matching. Journal of the American Medical Association 28: 3130.

[58] Ibid.

[59] Ibid 3131.

[60] Ibid 3133.

[61] Ibid. Haematopoietic reconstitution refers to the replenishment of blood cells by haematopoietic stem cells.

[62] Borger J, Meek J, 'Parents Create Baby to Save Sister', *The Guardian* (online) 4 October 2000 < https://www.theguardian.com/science/2000/oct/04/genetics.internationalnews > .

1.6 Future Uses for Embryo Testing

As discussed above, embryo testing is currently used to select against genetic disorders and positively select tissue matches. As the medical conditions for which testing is commonly utilised are single gene disorders, a test need only identify one gene. Further, apart from when testing is for the purpose of excluding susceptibility conditions such as cancer predisposition, doctors can say with a high degree of certainty that a specific gene will result in the future child developing the inherited condition. New tests are giving prospective parents the freedom to test for medical conditions or traits that are controlled by a combination of genes. For example, advances in genetics have led to the development of tests for polygenic conditions such as schizophrenia.[63] Furthermore, in the future, prospective parents may be able to test their embryos for intelligence, motivation, resilience or even immune capabilities. In the past, predicting the result of a particular combination of genes was the major technical barrier to the identification of polygenic traits in embryos.[64] However, it has been suggested that more accurate predictions are now becoming possible.[65] Through the use of genome-wide association studies, statistical models called 'predictors' can identify patterns in people's DNA.[66] Scientists can then determine the patterns that correspond to particular polygenic diseases or traits. It has been suggested that by analysing a person's DNA, the predictors can calculate their adult height to within three to four centimetres.[67]

1.7 Why Regulate Embryo Testing?

This section considers why embryo testing should be regulated and the goals that regulation should seek to achieve.

A common argument against the regulation of ART is that as natural conception is not regulated, ART should similarly be unregulated.[68] The argument goes that to regulate ART is therefore to discriminate against the infertile. This argument does not apply to embryo testing, as selecting which embryo to implant into a woman

---

[63] Tellier LCAM, Eccles J, Treff N et al. (2021) 'Embryo Screening for Polygenic Disease Risk: Recent Advances and Ethical Considerations'. Genes 12(8): 1105–1112.

[64] Shulman C, Bostrom N (2014) Embryo Selection for Cognitive Enhancement: Curiosity or Game-changer? Global Policy 5(1): 85.

[65] Regolado A, Eugenics 2.0: We're at the Dawn of Choosing Embryos by Health, Height and More (1 November 2017) MIT Technology Review <https://www.technologyreview.com/s/609204/eugenics-20-were-at-the-dawn-of-choosing-embryos-by-health-height-and-more/?source=download-meteredcontent>.

[66] Ibid.

[67] Ibid.

[68] See, eg, Douglas G (1991) Law, Fertility and Reproduction. Sweet & Maxwell, 111.

is not a choice available in the realm of natural conception.[69] Therefore, arguments based on the premise that natural conception is not regulated do not apply to embryo testing.

One reason to regulate embryo testing is to protect the future child's open future.[70] In the following chapters it will be argued that some uses of embryo testing, such as non-medical sex selection and selection for disability, deprive a child of an open future even before that child is born. Further, as it will be contended that providing a child with an open future is the key principle on which decision-making in this sphere should be based, it must therefore also be one of the principles to be protected through regulation. Regulation also means that sanctions can be imposed on clinics undertaking prohibited types of testing.

A further reason for the regulation of embryo testing is the promotion of high standards of clinical practice. At present, this is done through the reproductive technology accreditation committee, part of the Fertility Society of Australia and New Zealand.[71]

A final reason for regulating embryo testing is related to the dissemination of information and the gathering of data. One way this could be done is through the establishment of a regulator specialising in ART.[72] Such a regulator would be best placed to keep abreast of technological developments in the field of embryo testing and to communicate this information to both clinics and patients. Regulation would also enable the collection and analysis of data from clinics. Such data may include the kinds of technology utilised in the testing of embryos and its success rates. This information could also be made public to enable prospective patients to make informed choices about the clinic at which they receive treatment.[73]

The foregoing discussion suggests that there are three goals that the regulation of embryo testing can achieve. The first of these goals is the protection of the future child's open future. The second is the promotion of high standards of clinical practice. The final goal of regulation should be the sharing of information for the benefit of both patients and clinics. It appears that the current Australian regulatory framework for embryo testing is achieving the second of these goals. In the subsequent chapters,

[69] It is possible for a pregnant woman to 'select' their future child following natural conception, as they may choose to abort a foetus following prenatal diagnosis. For a good discussion of the law and ethics of prenatal diagnosis, see, eg, Karpin I, Savell K (2012) Perfecting Pregnancy. Cambridge University Press, Cambridge.

[70] Feinberg J (1980) The Child's Right to an Open Future. In: Aiken A, LaFollette H (eds) Whose Child? Children's Rights, Parental Authority and State Power. Littlefield, Adams & Co, 124.

[71] This is discussed in detail in chapter. 2.

[72] This is explored in detail in chapter. 8.

[73] Until 2021, the Victorian Assisted Reproductive Treatment Authority provided details about which clinics in Victoria carried out PGT and whether those tests were PGT-M, PGT-A or PGT-SR. After this time, clinics were only required to report their use of PGT-A. See Victorian Assisted Reproductive Treatment Authority <https://www.varta.org.au/resources/page/annual-reports>. General information about individual clinics' IVF success rates is available on the Your IVF Success website: https://yourivfsuccess.com.au

I will argue that changes to the regulatory framework may make it possible to achieve greater engagement with the first and third of these goals.

1.8 An Overview of the Book

In the film Gattaca, a geneticist advises prospective parents:

> You want to give your child the best possible start. Believe me, we have enough imperfection built in already. Your child doesn't need any additional burdens. Keep in mind, this child is still you. Simply, the best of you.[74]

Although a work of fiction, this quote reminds us how far ART has come since the birth of Louise Brown in 1978. ART is no longer solely a means by which infertile couples can have a genetically related child. It now raises questions about choice that go beyond the decision of whether or not to reproduce. Such questions lie at the heart of the regulation of embryo testing. This book examines the strengths and weaknesses of the regulatory framework for PGT and asks if there is a better way to regulate embryo testing in Australia. In proposing a revised regulatory framework, the book analyses the kinds of embryo testing currently permitted, as well as those prohibited under the current system. I argue that in addition to the two purposes for which embryo testing is currently permitted (PGT for the purposes of selection against disability and HLA typing), once the technology becomes available, the regulatory framework should also give prospective parents the freedom to use embryo testing for the purposes of selecting abilities in their future offspring. Further, I argue that the current prohibitions against non-medical sex selection and selection for disability should be maintained.

The book proposes a national, principles-based framework for the regulation of embryo testing. This framework would support clinics and prospective parents through the establishment of a national regulator, similar to the Human Fertilisation and Embryology Authority in the United Kingdom. Unlike the current Australian regulatory frameworks that set out limits on the kind of testing that may be undertaken, the proposed framework lays down broad principles to guide decision-making in this area.

This book presents a doctrinal analysis of the regulation of embryo testing in Australia. It engages with legal and bioethical research on PGT and HLA typing and takes an interdisciplinary approach, exploring areas such as disability studies and philosophy where relevant. The methodology used in this book is designed to explore the following two issues. First, what kind of children prospective parents should have the freedom to create through the use of embryo testing and selection, and second, how this freedom should be regulated.

---

[74] *Gattaca* (Directed by Andrew Nicol, Columbia Pictures, 1997).

1.8.1 Overview of Chapters

This section provides an overview of the forthcoming chapters of the book.

Chapter 2 provides an overview of the regulation of PGT and HLA typing in Australia. It examines the National Health and Medical Research Council Guidelines that apply across Australia, as well as the state legislation enacted in New South Wales, Victoria, South Australia, Western Australia, Queensland and the Australian Capital Territory. This chapter also looks at how the law relating to eligibility for treatment has developed over time, from the restrictive provisions of the 1980s that only permitted married, heterosexual couples access to treatment, to the current provisions that enable single women and lesbian couples to have a biologically related child. Consideration of the current regulatory framework for PGT and HLA typing highlights the problems inherent in the system (for example, inconsistencies between regulatory provisions) and thus the need for change.

Chapter 3 examines the United Kingdom's regulatory framework for PGT and HLA typing. The chapter charts the development of the law in this area from the Warnock Report to the *Human Fertilisation and Embryology Act 1990* (UK) and most recently the *Human Fertilisation and Embryology Act 2008* (UK). In addition to discussing the provisions relating to PGT and HLA typing, chap 3 looks at the reforms made to the 1990 Act by the 2008 Act, including amendments to the parentage provisions[75] and the welfare provision.[76] As will be seen, both the parentage provisions and the welfare provisions have developed to reflect the variety of family models that exist. This chapter also considers what Australia can learn from the UK and if there are any aspects of the UK framework that should be avoided in the development of a revised regulatory framework for Australia. The purpose of looking at the regulatory framework of an overseas jurisdiction is to outline an alternative model for regulation. The UK has been selected as it is an example of a national regulatory framework with a national regulator. Further, it has been described as a world leader in the fields of embryology and ART.[77] Australia and the UK also share a common legal heritage; thus it is interesting to examine the different ways that the regulation of embryo testing has developed in each jurisdiction. Regulatory frameworks in other countries are beyond the scope of this book.

Chapter 4 sets out the theoretical underpinnings of two concepts that play an important role in the forthcoming chapters. The chapter begins by asking, what is disability? This question is answered through the examination of various models of disability, including the medical, social, interactional, and welfarist models. Following consideration of each of these models, the interactional model is chosen as the most suitable model for use in discussions about reproductive choice. This chapter also introduces the term 'disabling condition' to ensure that discussions of

---

[75] These provisions govern who may be registered as the parents of a child born following ART.

[76] *Human Fertilisation and Embryology Act 2008* (UK) c 22, s 14(2)(b). This provision essentially serves the same purpose as eligibility criteria; it determines who may access ART.

[77] See, eg, United Kingdom, *Parliamentary Debates*, House of Lords, 13 September 2018, vol 792, col 2425 (Baroness Deech).

disability include genetic conditions such as cystic fibrosis, which may be viewed as an illness rather than disability. The second concept considered in this chapter is reproductive liberty. The chapter will consider the nature of the right to reproduce from differing perspectives before settling on an approach that can be applied in the subsequent chapters. The model of reproductive liberty that will ultimately be chosen is based on John Robertson's procreative liberty. Robertson's conception of procreative liberty is that of a negative claim right. By this he means that procreative liberty should not be interfered with without good reason. This model provides a good starting point from which to consider the scope of reproductive liberty, as it suggests that only harm to society or the future child are valid reasons to curtail prospective parents' reproductive liberty.

Chapter 5 considers the use of PGT to select against disability. In proposing a national framework for the regulation of embryo testing, two questions will be addressed. First, should parents have the freedom to test for and select against genetic conditions? Second, should there be limits on such testing? The interactional model of disability put forward in chap. 4 will form the backdrop for these discussions. The chapter considers the arguments advanced against the use of PGT for the purposes of selection against disability and asks whether any of these arguments represent a valid reason to prevent prospective parents using PGT. The arguments considered include the expressivist objection, the loss of support objection, and the claim that selection against disability is a eugenic practice. After concluding that these arguments are flawed and thus do not present adequate reasons to prevent prospective parents from using PGT, the discussion turns to the principles that should govern decision-making in this sphere. I suggest that two such principles are procreative beneficence and the open future argument. Finally, the chapter looks at the discrepancies between the legislative provisions operating in different states and the National Health and Medical Research Council Guidelines. One such discrepancy concerns the use of the word 'serious'. I ask whether this limit on the use of PGT is justified.

In chapter 6 I consider the future of embryo testing. The chapter begins with a discussion of whether prospective parents should be permitted to use PGT to select against polygenic conditions; conditions controlled by multiple genes. This technology is known as PGT-P. As many of the ethical issues around the use of PGT-P are the same as those discussed in the previous chapter, the discussion of PGT-P focuses on issues unique to this technology. The chapter then moves on to discuss testing embryos for traits that may be described as abilities. Within this chapter, the idea of general-purpose means is introduced. General-purpose means are traits that will be valuable to a person no matter what kind of life they choose to live. I will argue that embryo testing should be limited to traits that fall within the category of general-purpose means. Chapter 6 argues that the principle of procreative beneficence and the open future argument, which were both considered in chap. 5, also support giving prospective parents the freedom to select for ability characteristics in their future offspring, provided that these fall within the definition of general-purpose means. This chapter also illustrates the benefits of a principles-based system of regulation. As the principles in support of selection against disabling conditions also apply to selection for abilities, there does not seem to be a good reason for the regulatory framework to limit selection to the avoidance of disabling conditions.

Chapter 7 explores the conflict between reproductive liberty and the welfare of the child. This conflict is analysed by looking at three uses of embryo testing: selection for disabilities, non-medical sex selection and HLA typing. As discussed in Chapter 4, the model of reproductive liberty that will be applied is based on John Robertson's procreative liberty. In the context of selection for disabilities, it is argued that the conflict should be resolved in favour of the child. This argument is controversial as it is based on the idea of victimless harm. Nonetheless, deliberately selecting embryos that will be born with a disability, when other embryos are available, does not support the principle that a child's right to an open future should be protected.

The chapter then moves on to consider the conflict between reproductive liberty and the welfare of the child in the context of non-medical sex selection. As this kind of selection is motivated by gender stereotypes, it may also limit a child's right to an open future. It can therefore also be opposed on the grounds of the future child's welfare being the guiding principle. Lastly, HLA typing is discussed. I argue that HLA typing, both in conjunction with PGT and in its absence, should continue to be permitted, as there is no conclusive evidence of harm to the future child. That being said, counselling needs to be made available to parents and saviour siblings. This counselling should include the opportunity to meet others born under similar circumstances.

Chapter 8 considers the scholarship on principles-based regulation, with a view to constructing a national regulatory framework for embryo testing in Australia. The framework will permit the types of testing discussed in the preceding chapters, while prohibiting selection for disability and non-medical sex selection. A doctrinal analysis of principles-based regulation both in theory and in practice will be undertaken to determine its suitability for the proposed framework. I argue that this framework should have national application and thus the options for achieving this national regulatory reach are explored. The regulatory structure of the framework will also be discussed in this chapter. Within this discussion, a possible national regulator will be identified. This chapter concludes that a nationally applicable principles-based regulatory framework for embryo testing would allow Australia to achieve cohesion and consistency in the regulation of embryo testing, while also embracing technological innovation.

Chapter 2
The Regulation of Preimplantation Genetic Testing in Australia

Abstract This Chapter outlines the regulation of preimplantation genetic testing in Australia. It focuses on the regulation of tests for monogenic conditions and the selection of embryos that will be a tissue match for an existing person. It discusses both federal and state regulation and critiques the complexities that such a system presents. This chapter aims to provide a contextual background for discussions around reproductive choice that are the subject of the ensuing chapters.

2.1 Introduction

This chapter aims to give an overview of the regulation of preimplantation genetic testing (PGT) in Australia. The focus of this chapter will be the regulation of preimplantation genetic testing (PGT-M) for the avoidance of single gene disorders and human leukocyte antigen typing (HLA typing) for the selection of a tissue matched embryo. As the regulation of PGT-M and HLA typing in Australia comprises federal and state regulation, both will be considered in this chapter. Federal regulation exists in the form of the National Health and Medical Research Council's Ethical Guidelines[1] and the Fertility Society of Australia's Reproductive Technology Accreditation Committee, while state regulation varies significantly between states. Of the eight states and territories, two do not have legislation governing embryo testing or even assisted reproductive technology (ART) more generally.[2]

The chapter begins with an overview of the National Health and Medical Research Council Guidelines and the role of the Fertility Society of Australia. The chapter then moves on to consider the regulatory frameworks operating in the four states that

[1] National Health and Medical Research Council, *Ethical Guidelines on the Use of Assisted Reproductive Technology in Clinical Practice and Research 2017* ('*NHMRC Guidelines*').

[2] No legislation exists in Tasmania or the Northern Territory. No federal legislation exists in respect of ART due to Section 51 of the *Commonwealth of Australia Constitution Act 1901* (Imp). This section lists the areas in which the Commonwealth Parliament may legislate, with areas not covered by this section left to state Parliaments. Commonwealth legislation does, however, exist in respect of embryo research and human cloning. The relevant legislation is the *Research Involving Human Embryos Act 2002* (Cth) and the *Prohibition of Human Cloning Act 2002* (Cth).

M. de Souza, *The Regulation of Embryo Testing in Australia*,
https://doi.org/10.1007/978-981-96-5902-9_2

have legislation governing ART. The states are discussed in the order in which they enacted legislation. In addition to the regulatory provisions pertaining to PGT and HLA typing, the discussion of state regulation includes consideration of the way in which access to ART is regulated; because if access to ART is denied so is access to PGT. The chapter illustrates the problems with the Australian regulatory framework for embryo testing, and thus why a new framework should be developed.

2.2 National Health and Medical Research Council Guidelines

The National Health and Medical Research Council (NHMRC) is an independent statutory authority that sits within the portfolio of the Federal Minister for Health.[3] It was formed in 1936 and is the leading body for 'promoting the development and maintenance of public and individual health standards' in Australia.[4] The governing body of the NHMRC is the Council, which, among other things, provides advice to the chief executive officer and carries out any functions required of it by the Minister for Health.[5] The NHMRC has three core functions: it funds research on health and medicine, publishes health guidelines, and sets ethical standards.[6]

One of the NHMRC's principal committees is the Australian Health Ethics Committee (AHEC),[7] which issues guidelines on a variety of ethical issues relevant to health and human research. AHEC's key publication in the sphere of ART is the *Ethical Guidelines on the Use of Assisted Reproductive Technology in Clinical Practice and Research 2017* (NHMRC Guidelines). The purpose of the guidelines is to provide a comprehensive framework for the use of ART in clinical practice and research.[8] The guidelines have four parts: Part A introduces the guidelines, Part B provides a framework for the use of ART in clinical practice, Part C focuses on research pertaining to ART, such as research on human embryos and Part D covers mitochondrial donation.[9] The guidelines first came into being in 2004,[10] replacing

---

[3] National Health and Medical Research Council, *Leadership and Governance* < https://nhmrc.gov.au/about-us/leadership-and-governance/legislative-basis-nhmrc >. Accessed 15 April 2024.

[4] National Health and Medical Research Council, *Leadership and Governance* < https://nhmrc.gov.au/about-us/leadership-and-governance >. Accessed 15 April 2024.

[5] *National Health and Medical Research Council Act 1992* (Cth) s 21(1). Section 2.6 of the Act requires the NHMRC to have a CEO, who is the 'accountable authority' of the NHMRC.

[6] National Health and Medical Research Council, *Who We Are* < https://nhmrc.gov.au/about-us/who-we-are >. See also *National Health and Medical Research Council Act 1992* (Cth) s 3(1).

[7] *National Health and Medical Research Council Act 1992* (Cth) s 35. The Act also requires the NHMRC to have a Research Committee.

[8] National Health and Medical Research Council, *NHMRC Guidelines*, 13.

[9] Ibid 14–15.

[10] Ibid 15.

AHEC's 1996 publication, the *Ethical Guidelines on Assisted Reproductive Technology*.[11] In 2007, Part C of the guidelines was updated to reflect changes brought about by the *Prohibition of Human Cloning for Reproduction and the Regulation of Human Embryo Research Amendment Act 2006* (Cth). In 2017 Parts A and B of the guidelines were revised to ensure they remain relevant.[12]

The NHMRC Guidelines play an important role in the accreditation processes for ART clinics. Clinics are required to comply with the guidelines in order to gain accreditation and in this way the guidelines 'impose a duty of compliance.'[13] However, the courts have been eager to emphasise that the 'guidelines do not have the same status as a statute; and should not be interpreted or applied like a statute. The guidelines are intended to be just that—guidelines.'[14] Even so, if a clinic fails to comply with the NHMRC Guidelines they risk losing their accreditation, and operating without accreditation could lead to imprisonment.[15]

The NHMRC Guidelines set out a number of ethical principles for the use of ART in clinical practice, including the principle of respect for human embryos. The guidelines note that while there exist a number of different views about the moral status of the human embryo, one widely shared view is that the use of human embryos requires careful consideration.[16]

The NHMRC Guidelines address a variety of issues relevant to the practice of ART, including embryo testing,[17] consent,[18] counselling,[19] record keeping[20] and matters pertaining to the use of donated gametes.[21] The guidelines do not directly address the question of eligibility for treatment, instead leaving the decision up to individual clinics and state legislatures. They do, however, point out that decisions on issues of eligibility and access should be made openly, equitably and in a manner that is 'respectful of human dignity and the natural human rights of all persons'.[22] Further, the guidelines address the issue of access to treatment using the gametes of

---

[11] National Health and Medical Research Council, *Ethical Guidelines on the Use of Assisted Reproductive Technology in Clinical Practice and Research 2007*, 5 [1.1] ('*2007 Guidelines*').

[12] National Health and Medical Research Council, *NHMRC Guidelines*, 15. Following the addition of Part D in 2023, the official title of the Guidelines is the 'Ethical guidelines on the use of assisted reproductive technology in clinical practice and research 2017 (updated 2023)'.

[13] Skene L (2000) An Overview of Assisted Reproductive Technology Regulation in Australia and New Zealand. *Texas International Law Journal* 35: 31, 46.

[14] *YZ v Infertility Treatment Authority* [2005] VCAT 2655, [68] (Morris J).

[15] *Research Involving Human Embryos Act 2002* (Cth) s 11.

[16] National Health and Medical Research Council, *NHMRC Guidelines*, 21.

[17] Ibid 69–77.

[18] Ibid 37–39.

[19] Ibid 34–36.

[20] Ibid 83–85.

[21] Ibid 41–49.

[22] Ibid 20.

a deceased person and acknowledge the welfare implications for the intended child who will never know one of their parents.[23]

2.2.1 PGT

This section considers the provisions of the NHMRC Guidelines that relate to PGT. It discusses both the current guidelines and their 2007 predecessor. While it has been suggested that the NHMRC Guidelines 'do not provide a system of "regulation" for the use of PGD but set minimum standards'[24] in states that do not have legislation, or where the legislation does not cover PGT, the NHMRC Guidelines are an important regulatory tool. The guidelines permit the use of PGT, including sex selection,[25] for the purposes of testing for and selecting against 'genetic conditions, diseases or abnormalities that would severely limit the quality of life of the person who would be born'.[26] The guidelines put forward a number of criteria that clinics should consider when deciding whether or not to carry out testing, and note the importance of context when making such decisions.[27] Among the criteria that should be considered are evidence of the impact that the disorder will have on the future child, the ability of the family to care for a child with that condition and the availability of treatment to reduce the impact of the condition.[28] While the guidelines include reference to the experiences of individuals and their families who are living with the disabling condition for which PGT is sought, it is noteworthy that the first of the assessment criteria listed is 'current evidence and expert opinion on the impact of the condition'.[29] This may suggest that clinical opinion is to be considered before the experiences of individuals living with the condition.

The 2017 NHMRC Guidelines provided a long-awaited update to the NHMRC's 2007 guidelines. The 2007 guidelines prohibited the use of PGT for conditions that are not seriously harmful to the person to be born and for selection in favour of

---

[23] Ibid 79. Pursuant to the guidelines, clinics may only use gametes posthumously where the deceased person has left clear instructions consenting to their use (81 [8.22.1]). Where no instructions exist, the gametes of a deceased person may still be used, provided that they did not expressly object to this and certain conditions are satisfied ([8.22.2]–[8.22.3]). Clinics are also advised to seek guidance from an ethics committee, or other appropriate body, although the wording of the provision does not appear to make this a requirement ([8.23.1]).

[24] Smith M (2008) 'Reviewing Regulation of Assisted Reproductive Technology in New South Wales: The Assisted Reproductive Technology Act 2007 (NSW)' Journal of Law and Medicine 16: 120, 131. As noted in chapter 1, PGT-M was previously referred to as preimplantation genetic diagnosis (PGD) and therefore readers will notice references to PGD in pre-2017 literature.

[25] National Health and Medical Research Council, *NHMRC Guidelines*, [8.13.1].

[26] Ibid 73 [8.15.1].

[27] Ibid 74 [8.16.1].

[28] Ibid.

[29] Ibid.

disability.[30] Further, sex selection was only permitted for the purposes of reducing 'the risk of transmission of a serious genetic condition'. The 2007 guidelines did not define what counted as a serious genetic condition, but noted that this characterisation is controversial.[31]

The 2017 guidelines, like their predecessor, appear to permit PGT for late onset and 'lower penetrance' conditions,[32] with the age of onset being one of the factors to be taken into account when deciding whether or not PGT should be permitted.[33] The use of PGT-M for late onset conditions which in the future may be treatable or for lower penetrance conditions which may never develop, has been described as 'highly contentious' and 'a significant departure from traditional PGD'.[34] Jeanne Snelling, however, acknowledges that it may be misguided to place too much emphasis on penetrance, as the seriousness of the condition may be of equal importance.[35]

The NHMRC Guidelines acknowledge the ethical issues surrounding the use of PGT.[36] Among the issues considered by the NHMRC to be of ethical significance, are differing perceptions of disability, whether selecting against some disorders threatens 'the status of, and equality of opportunity for, people who have that condition', and the technical limitations of the technology. One of the notable omissions from the current guidelines is the removal of considerations pertaining to the disposal of healthy embryos as an ethical consideration.[37] This is to be welcomed, because rather than requiring clinics to take into account broad ethical issues, the criteria clinics are now directed to consider are more relevant to the situation in which prospective parents find themselves.[38] Kerry Petersen had criticised the previous iteration of the guidelines, on the basis that '[t]here is no requirement stipulating that a doctor must be *satisfied on reasonable grounds* that there is a risk of transmitting a serious genetic

---

[30] National Health and Medical Research Council, *2007 Guidelines*, 55 [12.2]. The prohibition on selection in favour of disability has been maintained in the 2017 Guidelines and can be found on page 73 [8.15.2].

[31] Ibid 55 [12.1]. In revising the guidelines, AHEC sought to avoid 'definitive statements about what constitutes a "serious" genetic condition', however, no such 'definitive statements' are found in the 2007 guidelines. See National Health and Medical Research Council, 'Summary of the Major Revisions to the ART Guidelines' (2017) 7.

[32] Snelling defines a 'lower penetrance' condition as 'a condition to which a person may be predisposed by virtue of a particular gene mutation they carry within their genetic code. "Penetrance" refers to the statistical likelihood of the disease occurring.' Snelling J (2008) 'Preimplantation Genetic Diagnosis for Susceptibility Conditions: A New Frontier or a Logical Extension?' Journal of Law and Medicine 16: 263, 263.

[33] National Health and Medical Research Council, *NHMRC Guidelines*, 74 [8.16.1].

[34] Snelling J (2008) 'Preimplantation Genetic Diagnosis for Susceptibility Conditions: A New Frontier or a Logical Extension?' Journal of Law and Medicine 16: 263, 264.

[35] Ibid 266. This is explored further in Chap. 5.

[36] National Health and Medical Research Council, *NHMRC Guidelines*, [8.16].

[37] National Health and Medical Research Council, *2007 Guidelines*, 55 [12.1]. The 2007 guidelines also required clinics to evaluate the ethical issues carefully, however, no such requirement is found in the current NHMRC Guidelines.

[38] See generally National Health and Medical Research Council, *NHMRC Guidelines*, 74 [8.16.1].

disorder or that a second medical opinion is necessary.'[39] The 2017 guidelines may be the target of similar criticism, although it is arguable whether such additional requirements are necessary, or even desirable.

The NHMRC Guidelines explicitly prohibit sex selection for non-medical purposes.[40] The guidelines do, however, leave the issue open by stating that the prohibition is dependent on further public debate.[41] The AHEC has noted that many different opinions exist on the issue of sex selection and suggest that there may be situations where sex selection is acceptable.[42] The majority of AHEC were of the view that there is an ethical difference between non-medical sex selection for the purposes of family balancing and non-medical sex selection purely for parental preference, with the latter being deemed unacceptable.[43] However, AHEC chose to maintain the prohibition on the grounds that non-medical sex selection is a controversial practice and further public debate needs to take place so that Australia is 'socially and politically' ready for the change.[44]

2.2.2 HLA Typing

The NHMRC Guidelines also address the issue of HLA typing and require clinics to seek advice from a clinical ethics committee, or other independent body, before testing embryos for tissue type.[45] In a move away from the 2007 guidelines that restricted the use of HLA typing to the creation of a tissue match for existing siblings,[46] the NHMRC Guidelines now also permit HLA typing in order to provide a stem cell transplant for parents and other relatives.[47] The justification for this change appears to be a recognition of the concept of 'biological relatedness', and the fact that harvesting stem cells from umbilical cord blood does not harm the future child.[48]

The guidelines set out key determinants that an ethics committee (or other body) should consider when deciding whether to permit HLA typing, including whether the future child will be a valued and 'respected member of the family'.[49] Ethics committees should also be satisfied that the use of HLA typing 'will not significantly

[39] Petersen K (2009) Regulating Preimplantation Genetic Diagnosis: A Criminal Model Versus a Professional Model. *Journal of Law and Medicine* 17: 452, 459.

[40] National Health and Medical Research Council, *NHMRC Guidelines*, 72 [8.14].

[41] Ibid 72.

[42] Ibid 71.

[43] Ibid 128.

[44] Ibid 127.

[45] National Health and Medical Research Council, *NHMRC Guidelines*, 75 [8.17.2].

[46] National Health and Medical Research Council, *2007 Guidelines*, 55 [12.3].

[47] National Health and Medical Research Council, *NHMRC Guidelines*, 74 [8.17].

[48] Ibid.

[49] Ibid 75 [8.17.2]. This appears to be an updated version of the 2007 provision that required ethics committees to consider whether parents actually wished to have another child, or if they were merely

affect the welfare and interests' of the future child.[50] In respect of the 2007 guidelines, Michelle Taylor-Sands argued that these requirements do 'not positively promote the importance [of] the child's interests over others'[51] and the same may be said of the current provisions. However, the introduction to the 2007 guidelines stated that, 'AHEC has recognised that the welfare of people who may be born as a result of the use of ART is paramount'[52] and it was arguable that therefore the requirements pertaining to HLA typing should be read in this light. Further, the 2007 guidelines also required that '[c]linical decisions must respect, primarily, the interests and welfare of the persons who may be born'.[53] This emphasis has been removed from the current NHMRC Guidelines and instead the emphasis appears to be on the balancing of the rights of the future child with those of the prospective parents[54] and in the context of HLA typing, other relatives.

2.3 *The Fertility Society of Australia and New Zealand*

In addition to the NHMRC Guidelines, the Fertility Society of Australia and New Zealand (FSANZ) also plays a role in the regulation of ART through its Reproductive Technology Accreditation Committee (RTAC). The FSANZ was established in 1982[55] and is a national body that represents the interests of those involved in reproductive medicine.[56] The objectives of the FSANZ include improving standards in reproductive medicine through the provision of clinical guidance and quality assurance, encouraging research into reproductive health, and providing its members with educational opportunities in order to raise practice standards.[57] Under the current regulatory framework, the RTAC, a sub-committee of the FSANZ, is responsible for the accreditation of ART clinics. The RTAC was established by the FSANZ in 1987,

---

looking for a source of tissue: National Health and Medical Research Council, *2007 Guidelines*, 56 [12.3.1].

[50] National Health and Medical Research Council, *NHMRC Guidelines*, 75 [8.17.2]. It is unclear why the current provision replaced the term 'adversely affect' with the more ambiguous 'significantly affect'. National Health and Medical Research Council, *2007 Guidelines*, 56 [12.3.1].

[51] Taylor-Sands M (2007) Selecting "Saviour Siblings": Reconsidering the Regulation in Australia of Pre-implantation Genetic Diagnosis in conjunction with Tissue Typing. Journal of Law and Medicine 14: 551, 561.

[52] National Health and Medical Research Council, *2007 Guidelines*, 9 [2.5].

[53] Ibid 21 [5.1].

[54] National Health and Medical Research Council, *NHMRC Guidelines*, 20.

[55] Fertility Society of Australia and New Zealand, https://www.fertilitysociety.com.au/professional-groups/pg-02/. Initially, the FSANZ was known as the Fertility Society of Australia.

[56] Fertility Society of Australia and New Zealand, About FSANZ https://www.fertilitysociety.com.au/about/about-01/. The FSANZ website states that it represents a wide range of people, from scientists and doctors to consumers and counsellors.

[57] Ibid.

a year after it produced the first set of standards for ART providers.[58] Following its establishment, the RTAC elaborated upon the original FSANZ standards. This set of standards formed the basis of the first RTAC Code of Practice. Clinics must comply with the RTAC Code of Practice in order to gain and thereafter retain accreditation.[59] The Code of Practice covers areas such as consent, the staff that must be employed by a clinic and their qualifications, and procedures for managing the risk of infection.[60] The RTAC Code of Practice does not, however, contain any ethical guidance for clinics, apart from stating that clinics need to comply with the NHMRC Guidelines.[61]

In addition to complying with the RTAC Code of Practice and the NHMRC Guidelines, to ensure they remain accredited clinics must also comply with any relevant legislation.[62] Failure to comply with these requirements will result in the loss of a clinic's RTAC licence. This in turn will have financial implications for clinics, as treatment[63] and pharmaceutical rebates are conditional upon RTAC accreditation.[64] Further, if a clinic continues to operate without accreditation, their staff may be imprisoned for 5 years.[65] Consequently, RTAC has been described as 'an important self-regulatory mechanism which ensures that the standard of treatment provided by Australian clinics is the responsibility of the professionals providing the treatment.'[66]

---

58 Reproductive Technology Accreditation Committee, *Code of Practice for Assisted Reproductive Technology Units* (October 2021) Fertility Society of Australia, 5 https://www.fertilitysociety.com.au/wp-content/uploads/20211124-RTAC-ANZ-COP.pdf Accessed 15 April 2024.

59 The *Research Involving Human Embryos Act 2002* (Cth) s 8 defines an accredited ART centre as a 'person or body accredited to carry out assisted reproductive technology by … the Reproductive Technology Accreditation Committee of the Fertility Society of Australia'.

60 Reproductive Technology Accreditation Committee, *Code of Practice for Assisted Reproductive Technology Units* (October 2024) Fertility Society of Australia https://fertilitysociety-com-au.membes-ams.com/public/267/files/RTAC/20241030-RTAC-ANZ-COP.pdf Accessed 7 February 2025.

61 Ibid [2.3(h)].

62 Ibid 10 [1.6].

63 Australian Government (2024) Medicare Services for Conceiving, Pregnancy and Birth https://www.servicesaustralia.gov.au/medicare-services-for-conceiving-pregnancy-and-birth?context=60092#a2.

64 Petersen K, Johnson M (2007) SmARTest Regulation? Comparing the Regulatory Structures for ART in the UK and Australia. Reproductive Biomedicine Online 15(2): 236, 239. As Petersen and Johnson point out, '[i]f patients had to bear the full burden of their treatment, they would most likely go elsewhere'. See also Pharmaceutical Benefits Scheme website < http://www.pbs.gov.au/browse/section100-if > .

65 *Research Involving Human Embryos Act 2002* (Cth) s 11. Prior to this legislation it had been suggested that should the market exist for prohibited applications of ART, 'there is little apart from economic restraint to prevent a clinic from forgoing RTAC accreditation': Szoke H, Neame L, Johnson L (2006) Old Technologies and New Challenges: Assisted Reproduction and its Regulation. In: Freckelton I, Petersen K (eds), *Disputes and Dilemmas in Health Law* (Federation Press) 187, 207.

66 Ibid 196.

2.4 State Regulation—Victoria

2.4.1 An Overview

The first generation of legislative responses to ART and indeed the first such piece of legislation in the world, was the *Infertility (Medical Procedures) Act 1984* (Vic). However, the Act was restrictive, contained a number of prohibitions and as John Leeton notes, 'reflected a strong religious bias against IVF'.[67] The statute was heavily criticised due to its perceived inability to keep pace with developments in ART and was considered a 'knee-jerk' legislative response.[68] Consequently, in little over a decade it was replaced by the *Infertility Treatment Act 1995* (Vic). The 1995 Act established the Infertility Treatment Authority,[69] a specialist regulator for ART in Victoria. The role of the Infertility Treatment Authority (ITA) was to grant licences to clinics for the practice of ART, to ensure clinics were complying with those licences, and to monitor the movement of gametes and embryos into and out of Victoria.[70] Further, the ITA was tasked with approving, overseeing and promoting research into the causes of infertility.[71] Under the framework created by the *Infertility Treatment Act 1995* (Vic), all treatment and research relating to ART had to be carried out pursuant to a licence in order for it to be legal.[72] Failure to abide by the ITA's licence conditions could result in a fine, a term of imprisonment of two years, or both.[73]

The *Infertility Treatment Act 1995* (Vic) was also not without its problems and like its predecessor was regarded as 'inflexible and incapable of keeping pace with new technologies and changing community views about families'.[74] However, this view was not universally shared, with some academics suggesting that the problems with the 1995 Act lay not in its inability to keep abreast of technological advancements, but rather in its 'prescriptive philosophy'.[75] This second generation legislative response soon came under review and in less than three decades the Victorian legislature passed its third ART statute.

On 1 January 2010, the *Assisted Reproductive Treatment Act 2008* (Vic) came into operation. This Act repealed the *Infertility Treatment Act 1995* (Vic) and is

---

[67] Leeton J (2004) The Early History of IVF in Australia and its contribution to the world (1970–1990). Australia and New Zealand Journal of Obstetrics and Gynaecology 44: 495, 498.

[68] Ibid.

[69] *Infertility Treatment Act 1995* (Vic) s 121.

[70] Ibid s 122(1).

[71] Ibid.

[72] Ibid ss 6, 22.

[73] Ibid 110.

[74] Thorpe R et al. (2011) New Assisted Reproductive Technology Laws in Victoria: A Genuine Overhaul or just Cut and Paste? Journal of Law and Medicine 18: 835, 847.

[75] Petersen K (2002) The Regulation of Assisted Reproductive Technology: A Comparative Study of Permissive and Prescriptive Laws and Policies. Journal of Law and Medicine 9: 483, 497.

considered 'less prescriptive' than its predecessors.[76] While the regulatory regime under the 2008 Act has been described as a 'criminal model', as breach of the statute may result in fines or imprisonment, it is also facilitative, permitting the use of ART within prescribed limits. The new Act brought with it a number of changes, including a new name to reflect 'the change in focus from treatment of infertility to a broader purpose of regulating assisted human fertilisation procedures.'[77] The change in focus brought about by the 2008 Act also brought with it a new set of guiding principles. However, the first principle (found in s 5(a) of the current Act), that 'the welfare and interests of any person born or to be born as a result of a treatment procedure are paramount', remained unchanged.[78] The second principle, namely that 'human life should be preserved and protected', has been removed and replaced with a non-commercialisation principle which[79] provides that ART should not be used for the commercial exploitation of women, men, or children born following treatment. The 1995 Act did not provide a definition of human life; however, given that one of the purposes of the Act was to regulate research on human embryos, it may be assumed that human life included the embryo in vitro.[80]

The final three guiding principles of the 2008 Act focus on access to information by children conceived through the use of donated gametes, the health of patients undergoing treatment procedures and non-discrimination.[81] In contrast, the final two guiding principles of the *Infertility Treatment Act 1995* (Vic), were that 'the interests of the family should be considered' and 'infertile couples should be assisted in their desire to have children'.[82] Further, the provision in s 5(2) of the 1995 Act, stating that the 'principles are listed in order of importance and must be applied in that order' has also been removed. This new set of principles would likely have been welcomed by commentators who felt that s 5(2) together with the four guiding principles did not provide an adequate framework in which to balance the 'reproductive interests of women with the contingent welfare interests of the resulting child.'[83] As Kerry Petersen noted in the context of the 1995 Act, 'it seems puzzling that the purported and speculative interests of eggs and sperm should take priority over the actual,

---

[76] Petersen K (2009) Regulating Preimplantation Genetic Diagnosis: A Criminal Model Versus a Professional Model. Journal of Law and Medicine 17: 452, 456.

[77] Victoria, *Parliamentary Debates*, Legislative Council, 10 October 2008, 4191 (Gavin Jennings).

[78] *Infertility Treatment Act 1995* (Vic) s 5(1)(a); *Assisted Reproductive Treatment Act 2008* (Vic) s 5(a).

[79] *Assisted Reproductive Treatment Act 2008* (Vic) s 5(b).

[80] Victoria, *Parliamentary Debates*, Legislative Assembly, 4 May 1995, 1245 (Marie Tehan). One possible reason for not including the principle of preservation and protection of human life in the 2008 Act, even in an amended form may be because the 2008 Act is more accepting of the reality that both ART treatment and research involve the destruction of human embryos. See Victoria, Parliamentary Debates, Legislative Assembly, 17 April 2007, 918 (James Merlino).

[81] *Assisted Reproductive Treatment Act 2008* (Vic) ss 5(c)–(e).

[82] *Infertility Treatment Act 1995* (Vic) ss 5(1)(c)–(d).

[83] Petersen K (2002) The Regulation of Assisted Reproductive Technology: A Comparative Study of Permissive and Prescriptive Laws and Policies. Journal of Law and Medicine 9: 483, 496.

verifiable interests of a living human being'.[84] Even in the absence of a provision equivalent to s 5(2) of the 1995 Act, it should be borne in mind that the operation of s 5(a) of the 2008 Act means that the most important consideration for decision makers must still be the welfare and interests of the future child.

The advent of the *Assisted Reproductive Treatment Act 2008* (Vic) brought with it a new specialist regulator for ART in Victoria, the Victorian Assisted Reproductive Treatment Authority (VARTA).[85] The functions of VARTA included administering the registration system for clinics, undertaking public education activities[86] and consulting the community on issues relevant to the 2008 Act.[87] Further, VARTA also played a role in the monitoring, promotion and review of matters relating to the Act.[88] In addition to establishing VARTA, the 2008 Act also established the Patient Review Panel.[89] The Patient Review Panel has a number of functions, including making decisions about surrogacy applications,[90] and considering cases where an ART provider has concerns about the welfare of a child who may be born following treatment.[91] The Panel is also responsible for considering applications for the posthumous use of gametes and embryos, and applications where a person does not meet the eligibility criteria set out in the Act.[92]

On 31 December 2024, VARTA ceased operation, and its functions were transferred to the Victorian Department of Health.[93] For many in Victoria, the loss of VARTA's specialist expertise in ART and its legislated public education function[94] will be a significant loss.[95,96]

[84] Ibid.

[85] *Assisted Reproductive Treatment Act 2008* (Vic) s 99.

[86] Ibid s 100(1)(b).

[87] Ibid s 100(1)(c).

[88] Ibid s 100(1)(d)–(f).

[89] Ibid s 82.

[90] Ibid s 85(a).

[91] Ibid s 85(d).

[92] Ibid s 85(c), (e).

[93] Victoria, Department of Health (2025) https://www.health.vic.gov.au/assisted-reproduction/assisted-reproductive-treatment-regulation Accessed 6 February 2025.

[94] *Health Legislation Amendment (Regulatory Reform) Act 2024* (Vic) s 74 substituted a new s 100 into the 2008 Act. This new section is solely focused on the donor conception register.

[95] See, eg, Clayton R (2024) Victoria's IVF regulator is internationally awarded for its fertility services—now it's facing the axe. https://www.abc.net.au/news/2024-07-09/victorian-ivf-fertility-treatment-authority-varta-sperm-donor/104053558 Accessed 6 February 2025.

[96] Petersen K (2009) Regulating Preimplantation Genetic Diagnosis: A Criminal Model Versus a Professional Model. Journal of Law and Medicine 17: 452, 453.

2.4.2 Eligibility for Treatment

This section considers the provisions governing who may access ART in Victoria. It begins by discussing the historical restrictions that were placed on access to assisted reproductive treatment, before moving on to consider the question of eligibility under the *Assisted Reproductive Treatment Act 2008* (Vic).

Both the *Infertility (Medical Procedures) Act 1984* (Vic) and its successor the *Infertility Treatment Act 1995* (Vic) limited access to ART to married and therefore heterosexual couples.[97] The couple also had to be infertile or at risk of having a child with 'a genetic abnormality or disease'.[98] In 1997, the marriage requirement was challenged by three de facto couples who were denied access to IVF under the 1995 Act. In *MW v Royal Women's Hospital*,[99] the Human Rights and Equal Opportunities Commission held the marriage requirement to be contrary to s 22 of the *Sex Discrimination Act 1984* (Cth). As this section makes it unlawful to discriminate against a person on the grounds of their sex or marital status in the provision of goods or services,[100] the hospital could not refuse to provide treatment on the grounds that the couples were not married.

A further legal challenge to the marriage requirement came in 2000. *McBain v Victoria*[101] concerned a doctor who found himself in a precarious predicament when presented with a single female patient seeking IVF treatment. If Dr McBain refused to treat his patient he would have been in breach of the *Sex Discrimination Act 1984* (Cth). On the other hand, providing her with IVF could have led to the loss of his licence, as he would have been acting contrary to s 8(1) of the *Infertility Treatment Act 1995* (Vic). Dr McBain therefore sought a declaration that s 8(1) of the 1995 Act was inconsistent with s 22 of the *Sex Discrimination Act 1984* (Cth). In the Federal Court, Sundberg J held s 8(1) to be invalid, pursuant to s 109 of the *Australian Constitution*.[102] The decision of Sundberg J was also the subject of an application for judicial review by the Australian Catholic Bishops Conference.[103] The High Court, however, unanimously refused to grant a writ of *certiorari* to quash the Federal Court's decision. Following *McBain*, the position under the 1995 Act was interpreted so as to permit clinics to provide fertility treatment to single women, but only if they were clinically infertile. Consequently, even after *McBain* the 1995

---

[97] *Infertility (Medical Procedures) Act 1984* (Vic) s 10(3)(a), *Infertility Treatment Act 1995* (Vic) s 8(1).

[98] *Infertility (Medical Procedures) Act 1984* (Vic) s 11(3)(d), *Infertility Treatment Act 1995* (Vic) s 8(3)(a)–(b).

[99] [1997] HREOCA 6.

[100] *Sex Discrimination Act 1984* (Cth) s 22(1).

[101] [2000] FCA 1009.

[102] *Australian Constitution* s 109 provides that '[w]hen a law of a State is inconsistent with a law of the Commonwealth, the latter shall prevail, and the former shall, to the extent of the inconsistency, be invalid'.

[103] *Re McBain; Ex parte Australian Catholic Bishops Conference* (2002) 209 CLR 372.

Act was still being construed in ways that were discriminatory, not only to single women but lesbian couples as well.[104]

In its 2007 report, *Assisted Reproductive Technology & Adoption*, the Victorian Law Reform Commission noted that 'the marital status requirement is not only inconsistent with the principle of non-discrimination, but also bears no relationship to the health and wellbeing of children, which must be the primary concern of the law governing ART.'[105] Consequently, the *Assisted Reproductive Treatment Act 2008* (Vic) does not contain any requirements pertaining to marital status, primarily because such provisions are unsustainable following the decision in *McBain*. The provisions of the 2008 Act also mean that a single or lesbian woman no longer needs to be clinically infertile to access treatment services.[106]

2.4.2.1 The Rise and Fall of the Presumption Against Treatment

The 2008 Act introduced a 'presumption against treatment' that prohibited clinics from treating a woman if a criminal record check revealed that the woman, or her partner (if applicable), had been convicted of a sexual or violent offence.[107] The Act also provided for a 'presumption against treatment' to be applied in situations where a child protection order[108] had been made and a child had been removed from the care of the woman or her partner.[109] Section 15 of the Act allowed an appeal to the Patient Review Panel for review of the decision not to provide the woman with treatment. Individuals also had the right to apply to the Victorian Civil and Administrative Tribunal for review of the Patient Review Panel's decisions.[110] This happened for the first time in 2011 in *ABY, ABZ v Patient Review Panel*.[111] The applicants were a couple who wished to use ART to conceive a child. The application to their ART provider included a criminal record check, which showed that in 2009 ABY had been convicted of sexual offences, namely three counts of 'sexual penetration of a 16 or 17 year old'[112] girl under his care. As a 'presumption against treatment' applied to the couple, the clinic was unable to provide treatment. ABY and ABZ then applied to

---

[104] Walker K (2002) The Bishops, The Doctor, His Patient and the Attorney-General: The Conclusion of the *McBain* Litigation. Federal Law Review 30: 507, 533.

[105] Victorian Law Reform Commission, *Assisted Reproductive Technology & Adoption: Final Report*, Report No 10 (2007) 67.

[106] *Assisted Reproductive Treatment Act 2008* (Vic) s 10(2)(a)(i).

[107] Ibid s 14(1)(a)(i)–(ii) as enacted.

[108] A child protection order check was required under *Assisted Reproductive Treatment Act 2008* (Vic) s 11(1)(d) as enacted.

[109] Ibid s 14(1)(b) as enacted.

[110] *Assisted Reproductive Treatment Act 2008* (Vic) s 96(a).

[111] [2011] VCAT 1382.

[112] *Sentencing Act 1991* (Vic) sch 1 cl 1(viiia).

the Patient Review Panel, which upheld the 'barrier to treatment'.[113] Consequently, the couple appealed to the Victorian Civil and Administrative Tribunal for review of the Patient Review Panel's decision.

The Tribunal considered a number of factors and concluded that no barrier to treatment should apply to ABY and ABZ. The factors considered by the Tribunal included ABY's character, the extent to which his status as a sex offender would enable him to participate in his child's life, and ABZ's character; the latter of which appeared to play a crucial role in the Tribunal's decision. Further, in considering ABY's motivations for his actions, the Tribunal accepted expert evidence that the sex of the victim rather than her age was the relevant motivating factor.[114] The Tribunal's decision was not, however, unanimous, with one member upholding the Patient Review Panel's decision to impose a barrier to treatment. Member Harper considered the seriousness of ABY's offence, his lack of empathy for his victim, his 'limited insight into his conduct' and the 'inconsistencies and contradictions in his evidence' as being risk factors that suggested a barrier to treatment should be imposed.[115] Furthermore, Member Harper felt that ABY's status as a registered sex offender had welfare implications for any child who may be born following ART.[116] The cautious approach adopted by Member Harper is understandable, especially considering the gravity of the consequences for the future child in the event of an error.

In October 2012, the Victorian Court of Appeal overturned the Tribunal's decision[117] on the grounds that the test applied by the Tribunal was too limited in its scope and not in line with the objectives of the 2008 Act. The Court held that the Tribunal should have applied a 'broader test' that considered 'any identifiable and established risk factors', not just those that gave rise to the presumption against treatment.[118] The case was returned to the Victorian Civil and Administrative Tribunal and re-heard by a Tribunal of different members in May 2013. The risk factors identified by the second Tribunal were firstly, the conduct for which ABY was sentenced,[119] secondly, ABY's denial that he had committed the offences[120] and thirdly, 'ABY's attitude and insight into his own conduct and the impact that attitude could have on the parenting of a child born to him and ABZ'.[121] The Tribunal also considered ABY's status as a registered sex offender and his history of substance abuse.[122] Considering each of

[113] *Assisted Reproductive Treatment Act 2008* (Vic) s 85(b) (as enacted) provided that one of the functions of the Patient Review Panel was 'to consider where there is a barrier to treatment if a presumption against treatment applies'.

[114] *ABY, ABZ v Patient Review Panel* [2011] VCAT 1382 [169].

[115] Ibid [286].

[116] Ibid [285].

[117] *Patient Review Panel v ABY & ABZ* [2012] VSCA 264.

[118] Ibid [118].

[119] *ABY & ABZ v Secretary to the Department of Health* [2013] VCAT 625 [75].

[120] Ibid [88].

[121] Ibid [105].

[122] Ibid [118], [123].

these risk factors in turn, the Tribunal concluded that none of the factors gave rise to a barrier to treatment.[123]

Between 2010 and 2020, 272 presumption against treatment applications were considered by the Patient Review Panel.[124] Whilst a barrier to treatment for those who fail the requisite checks seems prudent, Petersen argued that the 'barrier to treatment would appear to be a twenty-first century substitute for the former rules barring lesbian and single women from ART treatment.'[125] However instead of satisfying requirements pertaining to marital status, prospective parents had to pass criminal record and child protection checks. Further, Petersen noted that the restrictions 'were not supported by evidence in the parliamentary debates and are Orwellian in nature.'[126] In a similar vein, Sonia Allan has suggested that instead of imposing eligibility criteria, the focus should be on protecting children after they have been born.[127] The problem with this course of action is that in most instances it will require a child to suffer harm before steps are taken to address the problem. Therefore, the presumption against treatment and corresponding barrier to treatment had the potential to be important regulatory tools in the protection of the welfare of the child.[128] Further, rather than blanket restrictions, such as the requirements pertaining to marital status, the presumption against treatment and s 85(d) of the *Assisted Reproductive Treatment Act 2008* (Vic) provided for a two stage review process where a person's particular circumstances were considered in the first instance by the Patient Review Panel and if required by the Victorian Civil and Administrative Tribunal.

On 8 July 2020, the presumption against treatment was repealed by s 8 of the *Assisted Reproductive Treatment Amendment Act 2020* (Vic). One of reasons for revoking the provision was that in Victoria, all legislation must be compatible with the *Charter of Human Rights and Responsibilities Act 2006* (Vic). The presumption was deemed incompatible with the right to equality in s 8(2) of the Charter, on the basis that it indirectly discriminated against individuals on the basis of their sexual

---

[123] Ibid [128]. Other presumption against treatment cases appealed to the Victorian Civil and Administrative Tribunal include *TRV v Department of Health and Human Services* [2015] VCAT 1188, *CPA v Patient Review Panel (Human Rights)* [2016] VCAT 1555 and *OMU and RGJ v Patient Review Panel and Secretary to the Department of Health and Human Services (Human Rights)* [2018] VCAT 1235.

[124] Patient Review Panel, State Government of Victoria < https://www.vic.gov.au/patient-review-panel > Accessed 16 April 2024.

[125] Petersen K (2009) Regulating Preimplantation Genetic Diagnosis: A Criminal Model Versus a Professional Model. Journal of Law and Medicine 17: 452, 460.

[126] Ibid.

[127] Allan S (2012) Gatekeeping Parenthood: Should the State Play a Role in Determining Eligibility Criteria for Assisted Reproductive Treatment? Family Law Review 2: 149, 149.

[128] The presumption against treatment may also be opposed on the basis of the non-identity problem (discussed in Chap. 7), as it is based on the idea that in some cases it is better for a child not to exist at all – this has proved controversial. See, eg, Thompson K, McDougall R (2015) Restricting Access to ART on the Basis of Criminal Record: An Ethical Analysis of a State-Enforced "Presumption Against Treatment" With Regard to Assisted Reproductive Technologies. Journal of Bioethical Inquiry 12(3): 511; Karpin I (2012) The Legal and Relational Identity of the "Not Yet" Generation. Law, Innovation and Technology 4: 122, 124.

orientation or marital status.[129] For example, a same sex couple were perceived to be more affected by the requirements imposed by s 14 than a person in a heterosexual relationship, because they would need to use ART in order to conceive a child.[130]

A further reason given in Parliament for revoking the presumption was that 'the regulatory landscape for child safeguarding in Victoria has changed significantly' since the presumption came into operation, and thus the presumption was no longer required.[131] The problem with this argument is that measures related to child safeguarding all apply once the child has been born and harm has been done. For example, one of the measures referred to in Parliament was 'initiatives addressing family violence'[132]; such initiatives mainly deal with the fallout from violence as opposed to preventing harm in the first place. They are therefore not a replacement for the presumption, which was tasked with the function of preventing harm.

The most coherent of the arguments for revoking the presumption was that in the absence of the presumption the welfare of the future child would still be safeguarded through the guiding principles set out in s 5 of the *Assisted Reproductive Treatment Act 2008* (Vic).[133] As s 5(a) states that the 'welfare and interests of persons born or to be born as a result of treatment procedures are paramount', an ART provider may refuse treatment to a person if they reasonably believe that the person (or their partner) would pose a threat to their future child. This discretion to refuse treatment in the best interests of the future child has broader application than the presumption as it does not contain any defined restrictions. For example, in 2010, an ART provider refused to treat a single woman with a history of mental illness (a decision that was eventually overturned by the Victorian Civil and Administrative Tribunal).[134]

Prior to repealing the presumption, the Victorian government initiated an independent review of the regulatory framework for ART in Victoria. Chaired by lawyer Michael Gorton, *Helping Victorians Create Families with Assisted Reproductive Treatment: Final Report of the Independent Review of Assisted Reproductive Treatment* was published in May 2019. Recommendation 36 of the report was that the

---

[129] Victoria, *Parliamentary Debates*, Legislative Council, 19 March 2020, 1250 (Jaala Pulford).

[130] While no data was given to back up this argument, and such data does not appear to be collected in Victoria, the UK's Human Fertilisation and Embryology Authority reported that in 2021, 46,911 women with male partners underwent IVF, whereas only 2,201 women with female partners used IVF. If similar trends can be assumed in Australia, the indirect discrimination argument appears incorrect. Human Fertilisation and Embryology Authority (2023) Fertility Treatment 2021: Preliminary Trends and Figures https://www.hfea.gov.uk/about-us/publications/research-and-data/fertility-treatment-2021-preliminary-trends-and-figures/#section-7 Accessed 17 April 2024.

[131] Victoria, *Parliamentary Debates*, Legislative Council, 19 March 2020, 1252 (Jaala Pulford).

[132] Ibid.

[133] Victoria, *Parliamentary Debates*, Legislative Council, 19 March 2020, 1252 (Jaala Pulford).

[134] *PQ v Patient Review Panel* [2012] VCAT 291. It is worth noting that the Tribunal imposed a number of conditions that PQ had to satisfy, including monthly visits to a designated psychiatrist while she was undergoing treatment, followed by appointments every two months during pregnancy and for the first year following the birth of a child: [108]. NB: *Assisted Reproductive Treatment Act 2008* (Vic) s 91(3) gives the Patient Review Panel the power to impose conditions which 'it considers necessary and reasonable in the circumstances'. This power extends to the Victorian Civil and Administrative Tribunal when it is reviewing a decision of the Panel.

responsibility for arranging and sighting criminal record checks should fall to the ART provider, as opposed to a counsellor.[135] It suggested that shifting this responsibility would enhance the therapeutic relationship between the individual or couple seeking treatment and the fertility counsellor. The report did not recommend that the presumption be abolished.

During its period of operation, the presumption against treatment was controversial, and no doubt many people were happy to see it go. It remains to be seen whether this was the correct decision or whether implementing the recommendation of the Gorton Review would have been more desirable. As individuals and couples presenting for treatment are no longer required to undergo criminal record or child protection checks, the onus is now on doctors and ART providers in Victoria to determine whether treatment will be in the best interests of the future child. While this brings Victoria in line with other Australian jurisdictions, it is disappointing that the presumption was revoked without concrete evidence showing the absence of a link between past offences and harm to future children.[136]

2.4.3 PGT

This section considers the regulation of embryo testing in Victoria. It considers the current regulatory provisions, as well as the now repealed provisions of the *Infertility Treatment Act 1995* (Vic). Consideration of both of the Acts illustrates how far the legislation has come and where areas for improvement still exist. Neither the *Infertility Treatment Act 1995* (Vic) nor the *Assisted Reproductive Treatment Act 2008* (Vic) refer directly to PGT,[137] however, both Acts acknowledge the use of ART for the prevention of a genetic abnormality or disease in the embryo.[138] Neither statute defines 'genetic abnormality or genetic disease' and it has been suggested, in relation to the 1995 Act, that this was because 'Parliament intended the words to receive a broad, purposive interpretation'.[139] As a result, 'responsibility for gatekeeping' under the 1995 Act was imposed on the medical profession;[140] an approach

[135] Michael Gorton, *Helping Victorians Create Families with Assisted Reproductive Treatment: Final Report of the Independent Review of Assisted Reproductive Treatment* (May 2019) 60.

[136] De Souza M (2020) Should People with Criminal Records Be Able to Access Assisted Reproductive Technology? The Law in Victoria, Australia. BioNews 1064. Available via https://www.progress.org.uk/should-people-with-criminal-records-be-able-to-access-assisted-reproductive-technology-the-law-in-victoria-australia/

[137] Or PGD as it was previously termed.

[138] *Infertility Treatment Act 1995* (Vic) s 8(3)(b); *Assisted Reproductive Treatment Act 2008* (Vic) s 10(2)(c).

[139] Infertility Treatment Authority (Vic), 'Genetic Testing and the Requirements of the *Infertility Treatment Act 1995*: Policy in relation to the Use of Preimplantation Genetic Diagnosis' (2009) 4.

[140] Ibid.

that was criticised for disregarding 'subjective considerations in favour of "medical" or "scientific" criteria'.[141]

Under the 1995 Act, the ITA regulated PGT through a notification and approval system. Under this system, no approval or notification was required for conditions already approved by the ITA.[142] For conditions where testing had not previously been carried out, clinics needed to inform the ITA prior to carrying out the test.[143] In certain circumstances, case-by-case applications had to be made, including instances where PGT was used 'to select against carrier embryos, where the couple… is at risk of transmitting an autosomal recessive disorder', 'exclusion testing'[144] and where PGT was to be used in conjunction with HLA typing.[145] The ITA's policy on PGT had three main aims: first, facilitating the use of PGT for couples at risk of passing on a genetically linked or sex-linked disorder, secondly, to provide greater oversight for new applications of PGT and lastly, to prohibit certain uses of PGT.[146]

Under the 2008 Act, VARTA did not play an active role in the regulation of PGT, which is now managed by individual clinics, subject to the prohibitions in the legislation and provisions contained in the NHMRC Guidelines. This new 'hands off' approach to regulating PGT has been described as 'a radical change in policy'.[147] The only requirement from VARTA is that as part of the annual reporting process, clinics indicate whether they are offering PGT-A.[148] PGT-A has been singled out because, unlike PGT-M and PGT-SR, it is considered an add-on.[149]

The 2008 Act and its predecessor both prohibit the use of sex-selection for reasons other than the avoidance of a sex-linked disorder.[150] Section 28(2)(b) of the 2008 Act does, however, allow this prohibition to be circumvented subject to the approval of the Patient Review Panel. In 2010, a couple seeking to use PGT for the purposes of sex selection, in order to produce a female child, applied to the Patient Review Panel

---

141 Wagner EV, Mykitiuk R, Nisker J (2008) The "Affected" Post-Preimplantation Genetic Diagnosis Embryo. In: Bennett B, Carney T, Karpin I (eds) Brave New World of Health. Federation Press 37, 47.

142 Infertility Treatment Authority (2009) Genetic Testing and the Requirements of the *Infertility Treatment Act 1995*: Policy in relation to the Use of Preimplantation Genetic Diagnosis. [5.1].

143 Ibid.

144 Exclusion testing is the process by which an embryo is tested for an autosomal dominant condition while the affected parent can remain unaware of their own genetic status.

145 Infertility Treatment Authority (2009) Genetic Testing and the Requirements of the *Infertility Treatment Act 1995*: Policy in relation to the Use of Preimplantation Genetic Diagnosis, [5.4].

146 Ibid 2.

147 Smith M (2012) Regulating Assisted Reproductive Technologies in Victoria: The impact of changing policy concerning the accessibility of In Vitro Fertilisation for Preimplantation tissue-typing. *Journal of Law and Medicine* 19: 820, 834.

148 VARTA, 2023 Annual Report https://www.varta.org.au/resources/annual-reports Accessed 17 April 2024.

149 VARTA, 2023 Annual Report https://www.varta.org.au/resources/annual-reports Accessed 17 April 2024. Page 11 of the report states that PGT-M and PGT-SR are not considered add-ons because their use is support by robust evidence.

150 *Infertility Treatment Act 1995* (Vic) s 50, *Assisted Reproductive Treatment Act 2008* (Vic) s 28(1), (2)(a).

under s 28(2)(b) of the 2008 Act for permission to circumvent the prohibition in s 28. The couple had lost their only daughter at birth and already had three sons. There was no medical reason for the use of sex selection, as the couple were not at risk of transmitting a sex-linked disorder to the embryo. Since the death of their daughter the couple had suffered from post-traumatic stress disorder and some confusion existed at the Patient Review Panel hearing concerning whether a child of the desired sex was sought to assist their recovery.

The Panel held that if curing PTSD was the reason for seeking treatment, there was no evidentiary basis to support it.[151] The Panel also held that family balancing was 'not a sufficiently grave reason to approve a procedure that would otherwise be a criminal offence'.[152] As the Patient Review Panel rejected their application, the couple applied to the Victorian Civil and Administrative Tribunal for review of the Panel's decision.[153] The couple claimed that they were 'seeking sex selection in order to improve [their] emotional wellbeing' and to help them complete their family.[154] Further, JS explained that their desperation for a child was not as a replacement for the child who died, but rather 'to replace what was taken away from them'.[155] The Tribunal held that the various arguments advanced by the couple in support of sex selection focused on the needs of the couple as opposed to the welfare of the child to be born. Further, the argument that a sibling of the opposite sex would have a beneficial effect on the other children of the family,[156] was also construed as being about the existing children and not about the future child. As the 2008 Act requires the future child's welfare to be the paramount consideration,[157] the Tribunal dismissed the application for review of the Patient Review Panel's decision. Furthermore, the Tribunal stated that.

> if there is a conflict between the welfare and interests of a child to be born, and the health and wellbeing of the person undergoing an ART procedure, then, given the paramountcy of the welfare and interests of the child to be born under s 5(a), the conflict must be resolved in favour of the child's welfare and interests.[158]

The Tribunal's reasoning has been criticised by Smith and Taylor-Sands, who argue that the Tribunal did not clearly state why, in the particular circumstances of the case, non-medical sex selection and the welfare of the child principle were

---

[151] Cited in *JS and LS v Patient Review Panel* [2011] VCAT 856 [54].

[152] Ibid.

[153] *JS and LS v Patient Review Panel* [2011] VCAT 856.

[154] Ibid [53].

[155] Ibid [61].

[156] Ibid [81].

[157] Emily Jackson has argued that applying the welfare principle (which was developed for application to existing children in family law matters) to a child that has not yet been conceived 'is unjust, disingenuous and incoherent': Jackson E (2002) Conception and the Irrelevance of the Welfare Principle. Modern Law Review 65(2): 176–203, 181.

[158] *JS and LS v Patient Review Panel* [2011] VCAT 856 [87].

incompatible.[159] Further, turning to the Tribunal's reasoning relating to the couple's desire for a daughter being about the welfare of existing individuals, they note that having a child for the benefit of another person is permissible in cases of HLA typing.[160] Therefore, they argue that if HLA typing is permissible even though the reason for having a child is for the benefit of someone else, an argument based on the future child's welfare was not a valid basis for the Tribunal's dismissal of the application.[161] It is, however, important to note that HLA typing is only permitted under specific circumstances, for example, where treatment using stem cells is the recommended course of treatment.[162] Conversely, in *JS and LS v Patient Review Panel*, expert evidence indicated that JS's PTSD could in fact be made worse by a pregnancy.[163]

2.4.4 HLA Typing

As well as a policy on PGT, the ITA also published interim guidelines on the use of HLA typing in conjunction with PGT.[164] Neither the 1995 Act nor the ITA's interim guidelines addressed the issue of using HLA typing without PGT, in cases where the child to be born was not at risk of inheriting a genetic disorder, yet HLA typing was desired for the purpose of finding a tissue match for an existing sibling. However, the interim guidelines required prospective parents to provide an 'outline of the genetic condition to be tested for', suggesting that use of HLA typing in the absence of PGT was not permitted under the 1995 Act.[165]

The *Assisted Reproductive Treatment Act 2008* (Vic) permits the use of HLA typing in conjunction with PGT. In order to access HLA typing, individuals must satisfy the criteria in s 10 of the 2008 Act. This means that the woman, or her partner, must be at risk of passing on a genetic disease or disorder to a future child.[166] As

---

[159] Smith MK, Taylor-Sands M (2018) Comparing Non-Medical Sex Selection and Saviour Sibling Selection in the case of *JS and LS v Patient Review Panel*: Beyond the Welfare of the Child? Journal of Bioethical Inquiry 15: 139, 145.

[160] Ibid 149.

[161] Ibid.

[162] National Health and Medical Research Council, *NHMRC Guidelines*, [8.17].

[163] *JS and LS v Patient Review Panel* [2011] VCAT 856 [86].

[164] Infertility Treatment Authority (Vic), 'Tissue Typing in conjunction with Preimplantation Genetic Diagnosis: Interim Guidelines' (2007). The guidelines set out a number of key ethical considerations for deliberation by an ethics committee. These considerations, which appear inextricably linked, included 'the motivation and level of understanding of the parent in seeking to have an additional child', the issues that may arise if the tissue does not cure the sick child and 'the status of the child within the family' [2.3].

[165] Ibid [2.1].

[166] *Assisted Reproductive Treatment Act 2008* (Vic) s 10(2)(c).

Smith notes, HLA typing may then be undertaken as an ancillary procedure,[167] as it is not expressly prohibited under the Act. As access to treatment in Victoria is still dependent on the satisfaction of eligibility criteria, a couple would need to be at risk of passing on a 'genetic abnormality or genetic disease'; a requirement that cannot be satisfied where the condition suffered by the existing child is not inherited. However, s 10(1)(b)(ii) allows the criteria in s 10(2) to be circumvented subject to the approval of the Patient Review Panel. As one of the functions of the Patient Review Panel is 'to consider applications for treatment in circumstances in which the applicant does not meet the criteria for treatment',[168] it is likely that HLA typing without PGT would fall within this category.[169]

2.5 State Regulation—South Australia

2.5.1 An Overview

This section gives a brief overview of the regulation of ART in South Australia. ART in South Australia is governed by the *Assisted Reproductive Treatment Act 1988* (SA). The Act provides a framework for regulating ART that includes a system of registration for ART providers,[170] the establishment of a donor conception register[171] and provisions relating to confidentiality[172] and record keeping.[173] The impetus for introducing legislation to regulate ART came from a feeling that the rapid pace of scientific advancements had left society 'unprepared and uninformed to deal with the complex legal, social and ethical issues accompanying the developments.'[174] Further, it was argued that the medical and scientific professions should not be left to regulate themselves as 'it is undesirable and unreasonable to expect one part of society to shoulder such a burden.'[175] It was thus suggested that due to the 'scope and complexity of the issues',[176] parliamentary intervention was required.

---

[167] Smith M (2012) Regulating Assisted Reproductive Technologies in Victoria: The impact of Changing Policy Concerning the Accessibility of In Vitro Fertilisation for Preimplantation Tissue-typing. Journal of Law and Medicine 19: 820, 827.

[168] *Assisted Reproductive Treatment Act 2008* (Vic) s 85(1)(e).

[169] In Parliament, the creation of a saviour sibling was the example used to explain the operation of s 85(1)(e) of the 2008 Act. Victoria, *Parliamentary Debates*, Legislative Council 10 October 2008, 4192 (Gavin Jennings).

[170] *Assisted Reproductive Treatment Act 1988* (SA) Part 2.

[171] Ibid s 15.

[172] Ibid s 18.

[173] Ibid s 16.

[174] South Australia, *Parliamentary Debates*, Legislative Council, 10 September 1987, 872 (John Cornwall).

[175] Ibid 873.

[176] Ibid.

One of the key features of the legislation was the establishment of the South Australian Council on Reproductive Technology (SACRT). The remit of the SACRT extended beyond the scientific and medical sphere and addressed the 'broader issues' of morality, legality, 'questions of public policy, and most importantly, the welfare of the child.'[177] The SACRT was tasked with the development of a Code of Practice, licensing of clinics and promotion of public debate, among other things.[178] The wide ranging powers that were given to the SACRT generated significant debate in Parliament, with the leader of the Opposition arguing that many of the issues falling within the SACRT's remit should in fact be subjected to parliamentary scrutiny.[179] Despite arguments that the SACRT should be no more than an advisory committee,[180] the development of a 'code of ethical practice' for the regulation of ART remained one of its key functions.[181]

After two decades as the regulatory body for ART in South Australia, the SACRT was abolished by the *Reproductive Technology (Clinical Practices) (Miscellaneous) Amendment Act 2009* (SA) s 8. The reason for the abolition of the SACRT appears to lie in the notion that many of its functions are no longer required,[182] in part due to functions relating to the regulation of embryonic stem cell research now being dealt with under the *Research Involving Human Embryos Act 2003* (SA). The SACRT was replaced by the Ethics Health Advisory Council (EHAC), an advisory body covering a number of 'health-related ethical issues',[183] rather than specialising in ART. In addition to ART, included in EHAC's remit is organ donation, end of life issues and confidentiality. It had been suggested that EHAC would be supported by various subcommittees, one of which was to provide expert advice on issues related to ART.[184] While it was hoped that the SACRT's expertise would be harnessed within any newly formed ART advisory committee, this did not happen.[185] In her review of the South Australian regulatory framework, Sonia Allan points out that the EHAC has

[177] Ibid.

[178] Ibid.

[179] South Australia, *Parliamentary Debates*, Legislative Council, 20 October 1987, 1294 (Martin Cameron).

[180] South Australia, *Parliamentary Debates*, Legislative Council, 24 November 1987, 1938 (Rob Lucas).

[181] *Reproductive Technology (Clinical Practices) Act 1988* (SA) s 10(1)(a), repealed by *Reproductive Technology (Clinical Practices) (Miscellaneous) Amendment Act 2009* (SA) s 8.

[182] South Australia, *Parliamentary Debates*, House of Assembly, 26 November 2008, 1111 (John Hill).

[183] Government of South Australia, *Ethics Health Advisory Council Rules* (2010) [4.1].

[184] South Australia, *Parliamentary Debates*, House of Assembly, 26 November 2008, 1111 (John Hill). See also ibid [45] – [47].

[185] While the EHAC still appears to exist on paper, data from the Department of Health and Wellbeing (SA) suggests that in 2021, no members were appointed to the Council. Government of South Australia (2021) South Australian Government Boards and Committees Information https://www.dpc.sa.gov.au/responsibilities/boards-and-committees/reporting-board-membership-and-remuneration/South-Australian-Government-Boards-and-Committees-Information-30-June-2021.pdf Accessed 15 May 2024.

not fulfilled its role as a replacement for the SACRT 'to any great degree'.[186] Further, she notes that many South Australians 'are unaware of the EHAC's existence or functions'.[187] In her report, Allan recommends the establishment of an ART Advisory Council for South Australia.[188] She suggests the Council's role could include advising the Minister for Health on legal, moral, ethical and medical issues relevant to ART, as well as monitoring compliance with the 1988 Act, and public engagement on issues related to ART.[189]

2.5.2 *Eligibility for Treatment*

The original South Australian legislation, the *Reproductive Technology (Clinical Practices) Act 1988* (SA), limited access to ART to married or cohabiting couples.[190] In 1996, this was challenged in *Pearce v South Australian Health Commission*.[191] The plaintiff in *Pearce* was a single woman who wished to undergo IVF at a hospital operated by the defendant. The hospital declined to treat her on the basis that s 13(3) and s 13(4) of the *Reproductive Technology (Clinical Practices) Act 1988* (SA) only allowed married or cohabiting couples to receive treatment. The decision in *Pearce* was similar to that in *McBain*,[192] wherein the provision in the state legislation was held to be inconsistent with s 22 of the *Sex Discrimination Act 1984* (Cth).

Until 2017, the *Assisted Reproductive Treatment Act 1988* (SA) s 9(1)(c)(i) required a woman to be infertile in order to receive treatment. In the alternative, treatment could be provided 'if a man who is living with a woman (on a genuine domestic basis as her husband)' was infertile.[193] These provisions meant that while single and lesbian women were permitted access to treatment, they needed to satisfy either the requirement in s 9(1)(c)(i) or that in s 9(1)(c)(iii) detailed below. During the passage of the *Reproductive Technology (Clinical Practices) (Miscellaneous) Amendment Act 2009* (SA), it was argued in the Legislative Council that the amended legislation should contain provisions similar to those in Victoria, whereby a single or lesbian woman would not need to prove clinical infertility to have access to treatment.[194] It was suggested that this amendment would deter women from engaging

---

[186] Sonia Allan, 'Report on the Review of the Assisted Reproductive Treatment Act 1988 (SA)' (2017) 27.

[187] Ibid.

[188] Ibid 47.

[189] Ibid.

[190] *Reproductive Technology (Clinical Practices) Act 1988* (SA) s 13(3)(b), s 13(4).

[191] (1996) 66 SASR 486.

[192] *McBain v Victoria* [2000] FCA 1009.

[193] *Assisted Reproductive Treatment Act 1988* (SA) s 9(1)(c)(ii). This provision was repealed by the *Statutes Amendment (Surrogacy Eligibility) Act 2017* (SA) s 7(2).

[194] South Australia, *Parliamentary Debates*, Legislative Council, 8 September 2009, 3050 (Ian Hunter).

in interstate reproductive tourism or carrying out self-insemination with potentially unsafe sperm.[195] This amendment was, however, narrowly defeated by one vote.

The current iteration of the *Assisted Reproductive Treatment Act 1988* (SA) sets out a number of instances in which ART may be provided. Following the amendments brought about by the *Statutes Amendment (Surrogacy Eligibility) Act 2017* (SA), s 9(1)(c)(i) now provides that it must appear unlikely that a person seeking treatment will become pregnant other than through ART.[196] This means that single women and lesbians no longer have to be clinically infertile to access treatment. Treatment may also be provided where there is a risk of transmitting 'a serious genetic defect, serious disease or serious illness' to a child conceived naturally (s 9(1)(c)(iii)),[197] or subject to a number of conditions, in cases of posthumous conception (s 9(1)(c)(iv)).

2.5.3 PGT and HLA Typing

The *Assisted Reproductive Treatment Act 1988* (SA) does not contain any specific PGT provisions, however, s 9(1)(c)(iii) provides that one of the instances in which ART may be used, is 'if there appears to be a risk that a serious genetic defect, serious disease or serious illness would be transmitted to a child conceived naturally.' This suggests that access to treatment for the purpose of utilising PGT falls within the scope of the Act. Likewise, Petersen notes that although sex selection is not explicitly referred to in the legislation, 'sex selection to avoid the transmission of a sex-linked genetic defect clearly comes within the purpose of the Act.'[198] However, it could also be argued that at the time this provision was enacted, Parliament's intention was to facilitate the use of donor conception to avoid the transmission of a serious genetic condition.[199] Nevertheless, the wording of the provision suggests that the method of illness prevention was to be left open to take account of future developments. During its period of operation, SACRT considered the meaning of the term 'serious genetic defect' and concluded that it referred to 'a single gene or chromosome disorder the likely effect of which is to seriously impair a person who inherits the disorder'.[200]

---

[195] Ibid.

[196] *Statutes Amendment (Surrogacy Eligibility) Act 2017* (SA) s 7(2).

[197] In Parliament it was stated that the purpose of including the term 'serious disease' was to enable people with illnesses that are not genetic, such as those with HIV, to conceive a healthy child. South Australia, *Parliamentary Debates*, House of Assembly, 24 September 2009, 4141 (John Hill).

[198] Kerry Petersen, 'Genetic Technologies and ART: Ethical Values, Legal Regulation and Informal Regulation' in Ian Freckelton and Kerry Petersen (eds), *Disputes and Dilemmas in Health Law* (Federation Press, 2006) 217, 234.

[199] South Australia, *Parliamentary Debates*, Legislative Council, 20 October 1987, 1294 (Martin Cameron).

[200] South Australian Council on Reproductive Technology, 'Preimplantation Genetic Diagnosis (PGD) – Definition of "A Genetic Defect" for the purpose of Determining Eligibility for Treatment' (Memorandum No 12, 2006).

As HLA typing is not covered by the legislation, the NHMRC Guidelines provide the regulatory framework for the procedure in South Australia. However, individuals wishing to use HLA typing would still need to meet the eligibility criteria set out in s 9 of the 1988 Act. PGT with HLA typing would appear to be permitted in South Australia pursuant to s 9(1)(c)(iii) of the Act. The use of HLA typing where PGT is not required may be permitted under s 9(1)(c)(i) of the Act, as prospective parents may be able to argue that without access to ART they will not become pregnant as their 'circumstances' mean that they require a tissue match for their existing child.

2.6 State Regulation—Western Australia

2.6.1 An Overview

The *Human Reproductive Technology Act 1991* (WA) regulates the use of ART in Western Australia. During the passage of the Bill through Parliament, the reason for enacting legislation to regulate ART was that:

> the issues raised by reproductive technology and rapid advances in the treatment of infertility involve not only medical issues, but also scientific, social, religious, ethical and legal issues which go far beyond the bounds of traditional medical practice.[201]

Prior to the enactment of the legislation, it was reported that a clinic in Western Australia became the first to be denied national accreditation by the FSANZ.[202] It was therefore hoped that the legislation would provide 'a powerful but flexible approach to the regulation' of ART.[203] The Act was to achieve this by first, enforcing 'adherence to adequate standards and ethics, but in such a way that beneficial development [would not be] stifled'.[204] Second, by striking an 'appropriate balance between the rights of infertile couples… and the rights of any children born' and lastly, through the maintenance of 'equity, welfare and general standards in the community'.[205] The 1991 Act also established the Reproductive Technology Council (the Council),[206] which replaced the Interim Reproductive Technology Council. The Council's remit is

---

[201] Western Australia, *Parliamentary Debates*, Legislative Council, 13 June 1991, 3273 (Tom Stephens).

[202] Western Australia, *Parliamentary Debates*, Legislative Council, 13 June 1991, 3272 (Tom Stephens).

[203] Ibid.

[204] Western Australia, *Parliamentary Debates*, Legislative Council, 13 June 1991, 3273 (Tom Stephens).

[205] Western Australia, *Parliamentary Debates*, Legislative Council, 13 June 1991, 3274 (Tom Stephens).

[206] The Reproductive Technology Council was established by the *Human Reproductive Technology Act 1991* (WA) Pt 2.

to provide advice to the Minister for Health on matters related to ART and the 'administration and enforcement' of the Act.[207] The Council also has a role in licensing and the development and review of the Code of Practice.[208] Under the licensing system set up by the Act, clinics may be granted practice licences, storage licences, or licences that allow them to carry out both activities.[209] Western Australia is the only Australian jurisdiction to have a licensing system of this kind.

The guiding principles of the *Human Reproductive Technology Act 1991* (WA) differ from those in Victoria, as the 1991 Act does not contain a paramountcy provision; rather, one of the objectives of the Act is to ensure 'that the prospective welfare of any child to be born... is properly taken into consideration'.[210] However, s 4(1)(d) requires that the 'welfare of participants is properly promoted'.[211] These provisions appear to require clinics and the Reproductive Technology Council to undertake a balancing act, which may not necessarily result in the future child's welfare being treated as the primary consideration.

The 1991 Act also differs from its counterparts in Victoria and South Australia as it defines the term 'human embryo' and in doing so elucidates a starting point from which human life deserves protection.[212] While there were suggestions for the definition of human embryo to include an oocyte in the process of fertilisation,[213] these were rejected in favour of a definition similar to that currently found in the Act. When the Act was passed, the original provision referred to the 'completion of the fertilisation of the egg'.[214] In a similar vein, the definition of human embryo now contained in the Act is 'a live embryo that has human genome or an altered human genome and that has been developing for less than 8 weeks since the appearance of 2 pro-nuclei'.[215] This current definition and its predecessor are commendable for the following reasons; first, they set out to define a human embryo, as opposed to 'human life', and did this in a scientifically accurate fashion. Second, the provision identifies a point from which the embryo is given the protection of the law.

---

[207] *Human Reproductive Technology Act 1991* (WA) s 14(1)(a)(i)–(ii).

[208] Ibid s 14(1)(b)(i), (c).

[209] Ibid s 27.

[210] Ibid s 4(1)(d)(iv).

[211] Ibid s 4(1)(d)(iii).

[212] Ibid s 3A(1). This provision generated a great deal of debate in Parliament, because it was perceived as requiring the legislature to firstly, 'define the beginning of human life' and secondly, make a judgment on the sanctity of that life. Whether this was even the task at hand was debatable, with one member arguing that the aim was to define the stage at which 'an embryo is formed' as opposed to the considerably more monumental task of defining the beginning of life: Western Australia, *Parliamentary Debates*, Legislative Assembly, 14–16 May 1991. To further complicate the situation, it was noted that 'nowhere in our legal system have we yet defined the point at which life begins': Western Australia, *Parliamentary Debates*, Legislative Assembly, 16 May 1991, 2059 (Kevin Minson).

[213] Western Australia, *Parliamentary Debates*, Legislative Assembly, 16 May 1991, 2059 (Kevin Minson). This was supported by Graham Kierath in the same debate, 2066.

[214] *Human Reproductive Technology Act 1991* (WA) s 3(1).

[215] Ibid s 3A(1).

2.6.2 Eligibility for Treatment

Just like the early legislation in South Australia, the *Human Reproductive Technology Act 1991* (WA) limited access to ART to married and de facto couples who were infertile or at risk of transmitting a 'genetic abnormality or disease' to any child born.[216] In 2002, the position was amended by the *Acts Amendment (Lesbian and Gay Law Reform) Act 2002* (WA). The result of this Act was that single women and lesbians who were infertile or at risk of passing on a genetic abnormality to their offspring became eligible for treatment under the 1991 Act.[217] The statute now creates two classes of infertile women; those who are 'clinically infertile' and those who are 'socially infertile'.[218] While 'clinically infertile' women have access to ART regardless of their sexual orientation or marital status, 'socially infertile' women are left with limited reproductive choices. Further, the wording of s 23 excludes single men from accessing ART.

In Western Australia, treatment may also be refused if the reason for a woman's infertility is her age.[219] The effect of this provision is that post-menopausal women are not permitted access to ART in Western Australia. This criterion is not found elsewhere in Australia and therefore these women may travel interstate for treatment.[220] In 2018, the Western Australian government commissioned an independent review of the *Human Reproductive Technology Act* 1991 (WA) and the *Surrogacy Act 2008* (WA).[221] The review, completed in 2019, recommended greater clarity around the application and interpretation of the age restriction in s 23(1)(d),[222] and that further research and consultation be carried out regarding the appropriateness of the provision.[223] Further, the review recommended that discriminatory access provisions, such as those putting single men and clinically fertile single women outside the scope of the Act, be repealed.[224] In May 2022, the Minster for Health convened a Ministerial

---

[216] Ibid s 23(a), (c).

[217] Ibid s 23(1)(a)(ia), as inserted by *Acts Amendment (Lesbian and Gay Law Reform) Act 2002* (WA) s 74.

[218] Petersen K (2002) The Regulation of Assisted Reproductive Technology: A Comparative Study of Permissive and Prescriptive Laws and Policies. Journal of Law and Medicine 9: 483, 494.

[219] *Human Reproductive Technology Act 1991* (WA) s 23(1)(d).

[220] Unlike the plaintiffs in *McBain* and *Pearce*, an older woman would be unsuccessful in seeking a declaration of inconsistency with regard to s 23(1)(d) of the 1991 Act, as this restriction appears to be lawful pursuant to the *Age Discrimination Act 2004* (Cth) s 42(3).

[221] The review, carried out by Sonia Allan, made a total of 121 recommendations. It is published in two parts (Part 1 reviews the *Human Reproductive Technology Act 1991*) and is available via https://www.health.wa.gov.au/Articles/S_T/Submissions-for-HRT-review. Accessed 20 January 2025.

[222] Allan S (2019) The Review of the *Human Reproductive Technology Act 1991* and the *Surrogacy Act 2008* (Report: Part 1) recommendation 62 https://www.health.wa.gov.au/~/media/Files/Corporate/Reports-and-publications/HRT/Review-of-HRT-and-Surrogacy-Act-Part-1.pdf.

[223] Ibid recommendation 63.

[224] Allan S (2019) The Review of the *Human Reproductive Technology Act 1991* and the *Surrogacy Act 2008* (Report: Part 2) recommendation 3 https://www.health.wa.gov.au/~/media/Files/Corporate/Reports-and-publications/HRT/Review-of-HRT-and-Surrogacy-Act-Part-2.pdf.

Expert Panel that was tasked with providing advice on how new ART legislation could keep pace with scientific and medical developments, and social trends. The Panel put forward a number of recommendations, one of which would provide for gender-inclusive language.[225] With regard to eligibility for ART, the Panel proposed a set of broad criteria focused on necessity, the health of the future child and clinical discretion.[226]

2.6.3 *PGT and HLA Typing*

The 1991 Act permits clinics to undertake diagnostic procedures 'for the genetic testing' of embryos.[227] The Act requires all applications for such testing to be approved by the Council and prohibits the approval of diagnostic testing unless the procedure 'is unlikely to leave the embryo unfit' for implantation and 'there is a significant risk of a serious genetic abnormality or disease being present in the embryo'.[228] Although the Act does not define what counts as a 'significant risk' or 'a serious genetic abnormality or disease', guidance is provided by the Council. In respect of a 'significant risk', the Council has stated that '[i]t is not appropriate to specify a statistical probability as the sole criterion' on which to base decisions about significance.[229] Further,

> The level of risk should be measured against the risk of the disease or disability occurring in the general population. The Council should be satisfied that there is a higher risk of the embryo in question being affected… than for embryos in the general population.[230]

The Council's policy also acknowledges that the significance of a risk will depend on the perception of the person seeking treatment and clinics are advised to take these differing perceptions into account.[231]

With reference to the requirement that there be a 'serious genetic abnormality or disease' for which testing is being carried out, the Council advises that the seriousness of a genetic abnormality or disease depends on 'environmental and personal factors as well as the impairment to body functions and structures'.[232] This detailed guidance encourages transparency by providing clinics with a clear framework of the criteria used by the Council in its decision-making process. In early 2024, the

---

[225] Hart R (2022) Ministerial Expert Review Panel on Assisted Reproductive Technology and Surrogacy, recommendation 11 https://www.health.wa.gov.au/~/media/Corp/Documents/Health-for/ART/MEP-on-ART-and-Surrogacy-Final-Report.pdf.

[226] Ibid recommendation 10.

[227] *Human Reproductive Technology Act 1991* (WA) s 14(2b).

[228] *Human Reproductive Technology Act 1991* (WA) ss 14(2b)(a)(i)–(ii).

[229] Reproductive Technology Council, *Policy on Approval of Diagnostic Procedures Involving Embryos* (2017) 4 [5].

[230] Ibid 4 [6].

[231] Ibid 4 [7].

[232] Ibid 4 [8].

Council streamlined the approvals process for PGT by no longer requiring clinics to seek Council approval each time they needed to undertake embryo testing.[233] Clinics now need to check the PGT Approved Conditions list[234] and can use PGT for any of the conditions on the list without seeking further approval. If a condition is not on the list, clinics are required to submit an application to the Council. This application is considered by the Council's PGT Committee and if approved, the condition is added to the list of approved conditions.[235]

Neither the *Human Reproductive Technology Act 1991* (WA), nor the Council's policy on PGT provides any guidance on HLA typing. As the Act requires there to be 'a significant risk of a serious genetic abnormality… present in the embryo'[236] and the access requirements only permit access to IVF where a child is 'likely to be affected by a genetic abnormality or disease',[237] HLA typing may only be available where the embryo itself is at risk and not in situations where the aim of HLA typing is specifically to create a saviour sibling.[238] The Council has for some time considered this situation to be inequitable and has recommended changes to the Act[239]; however, these have not yet been made.

2.7 State Regulation—New South Wales

2.7.1 An Overview

Given that Victoria, South Australia and Western Australia passed ART statutes within 7 years of each other, it would be nearly two decades before ART legislation came into force in New South Wales. This is particularly puzzling in light of the

---

[233] Reproductive Technology Council, Preimplantation Genetic Testing https://www.rtc.org.au/preimplantation-genetic-testing/ Accessed 15 May 2024.

[234] Reproductive Technology Council (2024) Preimplantation Genetic Testing Approved Conditions List https://www.rtc.org.au/wp-content/uploads/2024/03/202403-PGT-approved-conditions-list-V3.pdf Accessed 15 May 2024.

[235] Reproductive Technology Council, Preimplantation Genetic Testing https://www.rtc.org.au/preimplantation-genetic-testing/ Accessed 15 May 2024. The factors considered by the PGT Committee when determining if a condition should be added to the approved conditions list are detailed in guidance from the Reproductive Technology Unit: Department of Health, Government of Western Australia (2025) Preimplantation Genetic Testing Applications to the Reproductive Technology Council: Guidelines for Clinics. https://www.rtc.org.au/wp-content/uploads/2025/03/20250328-Guidelines-for-clinics-REDCap-PGT-applications.pdf Accessed 24 July 2025.

[236] *Human Reproductive Technology Act 1991* (WA) s 14(2b)(a)(ii).

[237] Ibid s 23(1)(a)(ii).

[238] Taylor-Sands M (2007) Selecting "Saviour Siblings": Reconsidering the Regulation in Australia of Pre-implantation Genetic Diagnosis in conjunction with Tissue Typing. Journal of Law and Medicine 14: 551, 559. The usefulness of this distinction has been criticised, see Sheldon S, Wilkinson S (2004) Hashmi And Whitaker: An Unjustifiable And Misguided Distinction? 12 Med Law Rev 137.

[239] Reproductive Technology Council, *Annual Report 2008–2009*, 33.

fact that the New South Wales Law Reform Commission first reviewed the subject in 1986.[240] In 1997, the New South Wales Department of Health published a discussion paper on ART, as part of its review of the *Human Tissue Act 1983* (NSW).[241] The discussion paper looked at a range of aspects of assisted reproductive technologies, including eligibility for treatment, storage of gametes, record keeping, access to information and surrogacy and it was this paper that formed the basis of the *Assisted Reproductive Technology Act 2007* (NSW). In Parliament it was stated that the aim of the Act was 'to address a range of issues relating to the social and ethical aspects of assisted reproductive technology' that required a legislative response.[242] Further, the legislation was developed in line with three guiding principles: first, the recognition of 'obligations imposed on assisted reproductive technology providers by the existing laws'; second, acknowledgement of 'the rights of individuals to have control over the use of their genetic material'; and lastly, the principle that the best interests of the child are the paramount consideration.[243]

In 2010, the *Assisted Reproductive Technology Act 2007* (NSW) came into operation, with the guiding principles articulated in Parliament reflected in the objectives of the Act. Similar to the Victorian legislation, one of the aims of the 2007 Act is to 'prevent the commercialisation of human reproduction'.[244] Further, the Act seeks to 'protect the interests of… a person to be born as a result of ART treatment', gamete providers and women undergoing treatment.[245] The guiding principle that gamete donors should have control over their own genetic material has given rise to a curious provision not found in the legislation of the other states. Section 17(1) of the *Assisted Reproductive Technology Act 2007* (NSW) provides that 'A gamete provider may give an ART provider… written notice setting out the gamete provider's wishes in relation to the gamete'. The result of this provision is that as well as giving donors control over issues such as whether or not their gametes are stored or exported interstate, it also gives them the opportunity to discriminate against certain groups by expressly stating that they not be permitted to have access to their gametes.[246] Thus pursuant to this provision, a gamete donor could elect to prevent single or lesbian women, for example, from using their gametes. This provision generated significant debate in Parliament and it was argued that it 'grants legal sanction to bigotry and prejudice'.[247] Despite such objections, the provision was supported by a significant

---

[240] New South Wales Law Reform Commission, *Artificial Conception – Human Artificial Insemination*, Report No 49 (1986).

[241] New South Wales Department of Health, 'Review of the *Human Tissue Act 1983*: Assisted Reproductive Technologies' (Discussion Paper, 1997).

[242] New South Wales, *Parliamentary Debates*, Legislative Assembly, 7 November 2007, 3643 (Reba Meagher).

[243] Ibid.

[244] *Assisted Reproductive Technology Act 2007* (NSW) s 3(a).

[245] Ibid ss 3(b)(i)–(iii).

[246] Cf *Assisted Reproductive Treatment Act 2008* (Vic) s 17 which enables gamete donors to specify the kinds of procedures their gametes may be used in.

[247] New South Wales, *Parliamentary Debates*, Legislative Council, 28 November 2007, 4514 (John Kaye).

majority on the basis that it was in the best interests of the child not to find out later in life that one, or indeed both, of the gamete donors who helped make their existence possible in fact objected to their upbringing.[248]

The 2007 Act is novel as it is conceptually different from the legislation that exists in other Australian jurisdictions. The Act places a strong emphasis on the provision of information,[249] clinical standards[250] and record keeping.[251] Petersen has noted that rather than 'regulating most parts of ART practice, the New South Wales Act provides a broad framework for ART practice and aims to complement and enhance the present system.'[252] However, Smith argues that the exclusion of a number of issues, in order to avoid duplication of the NHMRC Guidelines, has led to the Act failing to provide a comprehensive system of regulation.[253] Further, he asserts that some provisions are effectively duplicates of those contained in the NHMRC Guidelines.

2.7.2 Eligibility for Treatment

The *Assisted Reproductive Technology Act 2007* (NSW) does not contain any eligibility criteria setting out who may access ART. This is interesting because if the purpose of the Act is to complement the existing framework, 'it would be expected that the legislation addresses some of the fundamental issues that are... not dealt with in any sense by the framework of professional and ethical guidelines'.[254] As both the NHMRC Guidelines and the RTAC Code of Practice are silent on the issue of eligibility for treatment, it is unclear why no guidance on this issue is found in the 2007 Act. The reason for the lack of eligibility criteria may be because Parliament felt it unnecessary to impose restrictions on who may access ART. Consequently, in a similar vein to the current Victorian and South Australian position, in New South Wales, single women and lesbians do not need to be clinically infertile to receive treatment. Therefore, the only barrier to access for these women would appear to be the ability of gamete donors to discriminate against them, pursuant to s 17(1) of the Act. The lack of eligibility criteria may give rise to reproductive tourism, whereby

---

248 New South Wales, *Parliamentary Debates*, Legislative Assembly, 7 November 2007, 3644 (Reba Meagher).

249 *Assisted Reproductive Technology Act 2007* (NSW) ss 13, 15.

250 Ibid ss 6–12, 16–29, 46–55.

251 Ibid ss 30–41 M.

252 Petersen K (2009) Regulating Preimplantation Genetic Diagnosis: A Criminal Model Versus a Professional Model. Journal of Law and Medicine 17: 452–461, 459.

253 Smith M (2008) Reviewing Regulation of Assisted Reproductive Technology in New South Wales: The Assisted Reproductive Technology Act 2007 (NSW). Journal of Law and Medicine 16: 120, 124.

254 Ibid 122.

individuals who have been denied treatment in other states may seek treatment in New South Wales.[255]

2.7.3 *PGT and HLA Typing*

The *Assisted Reproductive Technology Act 2007* (NSW) does not address the regulation of PGT or HLA typing and these issues remain regulated by the NHMRC Guidelines. In light of this, it is interesting to note that during the second reading of the Bill, reference was made to the legislation providing 'a strong regulatory framework for the ethical and social issues raised by [assisted reproductive] technologies'.[256] Given the range of ethical and social issues presented by PGT, some legislative guidance on this matter may have been expected. However, the decision to leave PGT out of the legislative framework in New South Wales suggests that Parliament considered the provisions of NHMRC Guidelines to be adequate. While some commentators have criticised the Act for failing to address the issue of PGT,[257] others have argued that the 'professional model' operating in New South Wales provides patients with 'more scope for reproductive choice'.[258]

The *Assisted Reproductive Technology Act 2007* (NSW) has been described as 'a "light touch" approach' to the regulation of ART.[259] Further, the lack of eligibility criteria and the decision not to establish a regulatory authority led to the legislation being described as 'the least prescriptive ART statute in Australia'.[260]

2.8 *State Regulation—The Australian Capital Territory*

The *Assisted Reproductive Technology Act 2024* (ACT) was passed by the Legislative Assembly for the Australian Capital Territory (ACT) in March 2024. The Act has two main objectives. First, the regulation of ART services, and second, the regulation of

---

[255] E.g. As noted at 2.6.2, older women and those deemed 'socially infertile' who do not satisfy the eligibility criteria for ART in Western Australia, may travel to NSW to access IVF.

[256] New South Wales, *Parliamentary Debates*, Legislative Council, 27 November 2007, 4382 (Tony Kelly).

[257] Smith M (2008) Reviewing Regulation of Assisted Reproductive Technology in New South Wales: The Assisted Reproductive Technology Act 2007 (NSW). Journal of Law and Medicine 16: 120, 131.

[258] Petersen K (2009) Regulating Preimplantation Genetic Diagnosis: A Criminal Model Versus a Professional Model. Journal of Law and Medicine 17: 452–461, 460.

[259] Petersen K (2013) Regulating Abortion and Pre-implantation Genetic Diagnosis in the Australian States of New South Wales, Victoria and Western Australia: Comparing and Contrasting Regulative Approaches. In: McLean S, Elliston S (eds), Regulating Pre-implantation Genetic Diagnosis: A Comparative and Theoretical Analysis. Routledge, Abingdon, p 126–147, 141.

[260] Ibid 142.

information associated with ART.[261] The Act's guiding principles include provisions protecting the welfare of the future child,[262] preventing exploitation[263] and giving donor conceived people 'a right to information about their donors'.[264] Further, in administering the Act, the health and wellbeing of those undergoing treatment 'must be protected'.[265] In line with its objectives, the 2024 Act sets up a regulatory framework for ART providers which closely aligns with that in New South Wales.[266] As the ACT is bordered by New South Wales on all sides, a framework created in this way assists in the delivery of health services across borders, by ensuring consistency and coherence.

In a similar vein to the *Assisted Reproductive Technology Act 2007* (NSW), the ACT legislation does not contain criteria setting out who may be eligible for ART. However, it is hoped that eligibility for treatment will be determined by clinics in a non-discriminatory manner and in line with the guiding principles of the NHMRC Guidelines.[267] As the *Assisted Reproductive Technology Act 2024* (ACT) does not contain specific provisions relating to PGT, nothing further will be said about it.

2.9 *State Regulation—Queensland*

2.9.1 An Overview

In September 2024, the Queensland Parliament passed the *Assisted Reproductive Technology Act 2024* (Qld). The Act establishes a regulatory framework for ART services in Queensland and a 'donor conception information register'.[268] The impetus for enacting legislation on ART came in 2023, when several incidents involving significant ART clinic failures gained media attention.[269] These incidents prompted the Minister for Health, Mental Health and Ambulance Services to direct the Health Ombudsman to undertake an investigation into ART services in Queensland.[270] On 28 March 2024, the Health Ombudsman released their interim findings. These findings

---

261 *Assisted Reproductive Technology Act 2024* (ACT) s 7.

262 Ibid s 8(a).

263 Ibid s 8(b).

264 Ibid s 8(c).

265 Ibid s 8(d).

266 Explanatory Statement, Assisted Reproductive Technology Bill 2023 (ACT) 1.

267 National Health and Medical Research Council, *NHMRC Guidelines*, 15.

268 *Assisted Reproductive Technology Act 2024* (Qld) Part 3.

269 See, eg, Lackey B (2023) The Dad with 1,000 Kids: Why this 'broken' mother-of-five fears her children will be forced to run background checks on every potential love interest—or face an unthinkable reality—as an entire state is put on alert. Daily Mail Australia https://www.dailymail.co.uk/news/article-12708721/Sperm-donor-1000-kids-Queensland-mother-five-fears-children-forced-run-background-checks-love-interests.html Accessed 27 May 2024.

270 This investigation was conducted pursuant to s 81 of the *Health Ombudsman Act 2013* (Qld).

suggested that self-regulation in Queensland was problematic and a robust regulatory framework was required.[271] The regulatory framework that the Act seeks to establish aims to introduce 'greater oversight, transparency and safeguards' into the provision of ART services, thus improving confidence in the delivery of ART treatment in Queensland.[272]

The objects of the *Assisted Reproductive Technology Act 2024* (Qld) include protecting 'the welfare and interests' of those using ART and children born through ART, the regulation of ART, and the provision of information to donor conceived individuals.[273] Like other Australian ART statutes, the interests of those born following an ART procedure are the paramount consideration in the administration of the Act.[274] Admirably, the Act avoids the use of gendered terminology such as woman and instead uses the term person.[275]

2.9.2 Eligibility for Treatment

The *Assisted Reproductive Technology Act 2024* (Qld) does not address eligibility for treatment. This may have been a missed opportunity given the history of this issue in Queensland. In *JM v QFG*,[276] the Queensland Court of Appeal held that a clinic's decision not to treat a lesbian couple was not discriminatory, on the basis that the woman seeking treatment was not clinically infertile or at risk of passing on a genetic defect to any child born. Therefore, the decision not to treat the woman was based on medical reasons and not on her sexual orientation.[277] This case illustrates that a lack of legislative guidance can result in a lack of transparency for patients and exposes clinicians to unnecessary litigation.

2.10 State Regulation—Tasmania and the Northern Territory

In Tasmania and the Northern Territory no legislation exists to regulate ART. In these jurisdictions, the NHMRC Guidelines form the basis of medical decision-making. As the NHMRC Guidelines are silent on the subject of eligibility for treatment, one may assume that clinics in these states would not impose restrictions on who may receive treatment.

---

[271] Explanatory Notes, Assisted Reproductive Technology Bill 2024 (Qld), 3.

[272] Ibid 4.

[273] *Assisted Reproductive Technology Act 2024* (Qld) s 3(1).

[274] Ibid s 3(2).

[275] E.g. s 5(1)(a) of the 2024 Act refers to attempts procure a 'pregnancy in a person'.

[276] [2000] Qd R 373.

[277] Ibid 386.

As well as reliance on the NHMRC Guidelines, clinics in the Northern Territory follow the South Australian system for the regulation of ART. This means that they are required to comply with the *Assisted Reproductive Treatment Act 1988* (SA), except where it conflicts with any Northern Territory legislation.

Two main reasons may be given for why these states and territories should enact legislation. First, a lack of legislation may lead to inconsistencies within a state in relation to issues not covered in the NHMRC guidelines.[278] Secondly, 'statutory regulation has a broader interest' than, for example, professional self-regulation and has been described as 'the process of a democratic legislature providing a framework which expresses the protection of the public interest'.[279] Consequently, the moral and ethical issues raised by ART should be debated in parliament to ensure any resulting regulation best represents the public interest.

2.11 Conclusion

This chapter has given an overview of the regulatory framework for embryo testing in Australia. It has considered the way in which PGT and HLA typing is regulated, as well as the policies governing access to these technologies. Table 2.1 summarises the relevant legislative provisions. The strengths of the Australian system of regulation for PGT and HLA typing lie in its facilitative framework and comprehensive NHMRC Guidelines. However, the mixture of national guidelines, a national accreditation committee and a variety of state legislation and regulators has led to a complex, multi-tiered regulatory model.

The major weakness of the Australian regulatory framework is the lack of consistency that invariably results from a multi-tiered system. For example, the differing eligibility criteria governing who may access ART means single women and lesbians must be clinically infertile to access treatment in Western Australia, whereas in South Australia, New South Wales and Victoria no such restrictions apply. This may give rise to interstate reproductive tourism, whereby women unable to access treatment in one state travel to another state where they satisfy the eligibility criteria. A further example of the lack of consistency across the states and territories can be found in the terminology used with regard to PGT. The NHMRC Guidelines refer to the use of PGT for the avoidance of a genetic condition, abnormality or disease 'that would severely limit the quality of life of the person who would be born'; however, the Victorian legislation simply refers to 'a genetic abnormality or genetic disease'. The South Australian statute, on the other hand, makes reference to a 'serious genetic defect', whereas the Western Australian Act requires there to be 'a significant risk of a serious

[278] Smith M (2007) 'Revisiting Old Ground in Light of New Dilemmas: The Need for Queensland to Reconsider the Regulation of Assisted Reproductive Technologies' Queensland University of Technology Law and Justice Journal 7(2): 425, 426.

[279] Szoke H (2002) 'The Nanny State or Responsible Government' Journal of Law and Medicine 9: 470, 477.

Table 2.1 Summary of ART statutes in Australia

	Legislation	Eligibility Criteria	Provisions governing PGT and HLA Typing	Regulator specialising in ART
Victoria	*Assisted Reproductive Treatment Act 2008* (Vic)	s 10	s 10(2)(c), s 10(3), s 28	Secretary for the Department of Health
SA	*Assisted Reproductive Treatment Act 1988* (SA)	s 9(1)(c)	s 9(1)(c)(iii)	None
WA	*Human Reproductive Technology Act 1991* (WA)	s 23(1)	s 14(2b)(a)	Reproductive Technology Council
NSW	*Assisted Reproductive Technology Act 2007* (NSW)	None	None	None
Queensland	*Assisted Reproductive Technology Act 2024* (Qld)	None	s 24	None
ACT	*Assisted Reproductive Technology Act 2024* (ACT)	None	None	None
Tasmania	None	N/A	N/A	N/A
NT	None	N/A	N/A	N/A

genetic abnormality or disease'. This differing terminology is unhelpful as it raises questions such as, is a genetic defect the same as a genetic abnormality? Or, how does a genetic disease differ from a genetic condition? The provisions that bear the most similarity to each other are those found in the Victorian and Western Australian legislation. However, even reconciling these provisions may prove complicated, as it may be questioned whether the reference to 'risk' in the *Assisted Reproductive Technology Act 2008* (Vic) has a different meaning to 'significant risk' in the *Human Reproductive Technology Act 1991* (WA).

A final example of this lack of consistency stems from the use of HLA typing without PGT. In Western Australia this is not permitted. While the situation in South Australia is unclear, it is possible that HLA typing in the absence of PGT may be approved. As the NHMRC Guidelines do not distinguish between cases where PGT is used in conjunction with HLA typing and cases where only HLA typing is required, in states that rely solely on the Guidelines, the use of HLA typing to determine tissue compatibility is permitted even in cases where the sick relative does not suffer from a genetic condition.

One possible way to address the inconsistencies that are present in the current model would be to have federal legislation rather than state legislation governing embryo testing, or perhaps ART more generally. In a similar vein to the process that resulted in the *Research Involving Human Embryos Act 2002* (Cth), legislation could be enacted pursuant to the corporations power in s 51(xx) of the *Commonwealth of Australia Constitution Act 1901* (Imp).[280] Further, in light of the enactment of the first ART statute in Victoria in 1984, it is noteworthy that 40 years later Queensland and the ACT have both finally passed legislation on ART. Over the last 40 years, most of the states which have passed legislation have updated their statutes to take account of evolving technologies and societal changes, and this process would be much more efficient if it was carried out at Federal level. Federal legislation would also mean that states without legislation would be brought under a national framework, rather than passing reactionary statutes following adverse events; as was the case in Queensland.

The regulation of PGT and HLA typing in Australia requires further consideration, as there are a number of issues that warrant further investigation. For example, is permitting the use of PGT for lower penetrance conditions acceptable? Further, in states where there is either no regulatory body, or where the regulatory body adopts a 'hands off' approach, decisions regarding which conditions may be tested for using PGT are left to clinics. This gives rise to the question, are doctors best placed to make this decision? Additionally, does the current Australian regulatory framework place acceptable limitations on reproductive choice? In the chapters that follow, these questions will be among those considered in the development of a national framework for the regulation of PGT and HLA typing in Australia.

---

[280] This is discussed in detail in Chap. 8.

Chapter 3
The Regulation of Preimplantation Genetic Testng in the United Kingdom

Abstract This chapter provides an overview of the regulatory framework for embryo testing in the UK and compares this to the Australian regulatory model for embryo testing described in the previous chapter. The UK has been chosen as a comparator because it is considered a world leader in the regulation of assisted reproductive technology. As a result, this chapter considers whether there are useful aspects of the UK system of regulation that could play a role in the development of a revised regulatory framework for embryo testing in Australia.

3.1 Introduction

This chapter considers the regulatory framework for embryo testing in the United Kingdom. The United Kingdom (UK) has a comprehensive framework for the regulation of assisted reproductive technology (ART) and is considered a world leader in this field. Thus, the purpose of this chapter is to establish whether there is anything Australia can learn from the UK when developing a revised framework for the regulation of embryo testing. The chapter begins by looking at the history of ART regulation in the UK, which started with the Report of the Committee of Inquiry into Human Fertilisation and Embryology (the Warnock Report). The Warnock Report, named after the Chair of the committee, the late Baroness Mary Warnock, has been described as 'one of the most influential, pragmatic and farsighted reports of its kind'.[1] The Report was published in 1984, however, it was not until 1990 that the *Human Fertilisation and Embryology Act*[2] was enacted.[3] This Act created the Human Fertilisation

[1] Deech R, Smajdor A (2007) From IVF to Immortality. Oxford University Press 8.

[2] *Human Fertilisation and Embryology Act 1990* (UK).

[3] It has been suggested that this delay was due to 'deliberate stalling tactics by pro-embryo research factions within the Government', thus giving supporters of embryo research extra time to win over those MPs who were initially sceptical. See Snelling J, Gavaghan C (2015) PGD Past and Present: Is the HFE Act 1990 Now 'Fit for Purpose'? In: Horsey K (ed), *Revisiting the Regulation of Human Fertilisation and Embryology* (Routledge) 80, 82.

M. de Souza, *The Regulation of Embryo Testing in Australia*,
https://doi.org/10.1007/978-981-96-5902-9_3

and Embryology Authority (HFEA),[4] which has been described as the 'envy and blueprint for the world'.[5]

After providing a brief overview of the Act and the HFEA, the chapter then moves on to consider the way in which eligibility for treatment is regulated in the UK, including the controversial welfare provision. Following this, specific issues pertaining to the regulation of PGT will be discussed. PGT-M for the purposes of selecting against disability will be considered first. The discussion will then move on to positive selection. The only kind of positive selection currently permitted in the UK, human leukocyte antigen typing (HLA typing) to select a tissue matched embryo, will be examined. Analysis will then turn to non-medical sex selection and selection in favour of disability, both of which are prohibited under the 1990 Act. The chapter will end with a comparative analysis of the regulation of embryo testing in both the UK and Australia. This comparative analysis will then be used to inform recommendations for the development of a more effective regulatory framework for embryo testing in Australia.

3.2 Overview

In July 1982, four years after the birth of Louise Brown, the UK government formed a committee to investigate the 'social, ethical and legal implications' of developments in assisted reproduction.[6] The Committee of Inquiry into Human Fertilisation and Embryology, chaired by the philosopher Dame Mary Warnock, comprised medical professionals, lawyers, social workers and academics. The committee produced a comprehensive report that was submitted to Parliament in June 1984. The report covered artificial insemination, IVF, gamete and embryo donation, storage of gametes and embryos, regulation of treatment services, and the regulation of research. The committee made a number of recommendations, including that a statutory licensing authority should be established to regulate treatment services and research.[7] It was suggested that the body should have a significant number of members from outside the medical and scientific professions, including a lay chair. This varied membership would prevent the perception that the body was 'unduly influenced by sectional interests'.[8] The committee felt that this was important in order to ensure public confidence in the proposed body.[9] They envisioned the body as having two functions; an advisory function and an 'executive' function. In its advisory function, the body would issue

[4] *Human Fertilisation and Embryology Act 1990* (UK) s 5.

[5] United Kingdom, *Parliamentary Debates*, House of Lords, 1 February 2011, vol 724, col 330 (Baroness Thornton).

[6] Department of Health and Social Security (UK), *Report of the Committee of Inquiry into Human Fertilisation and Embryology* (Cmd 9314, 1984) iv ('Warnock Report').

[7] Ibid 80.

[8] Ibid 76.

[9] Ibid 75–76.

guidance to clinics and researchers, as well as providing advice to the government when required.[10] In its executive function, the body would grant treatment licences to clinics providing ART services, and research licences to scientists undertaking researching involving human gametes and embryos.[11]

A further notable recommendation of the committee was the '14-day rule',[12] which still forms part of the law today.[13] The Warnock Report recommended that no human embryo should be allowed to develop in vitro for longer than fourteen days; one day before the appearance of the primitive streak.[14] Therefore, by this point, any embryos used in treatment must be transferred to a woman or placed in storage, and any research being carried out on human embryos must cease. In recent times, the 14-day rule has been the subject of much debate, with many calling for this limit to be extended.[15]

3.2.1 From the VLA to the HFEA

Following a hostile reception towards the Warnock Report in the House of Lords, in 1985 the Royal College of Obstetricians and Gynaecologists and the Medical Research Council established the Voluntary Licensing Authority (VLA). The heads of both organisations hoped the establishment of a licensing body for assisted reproduction would help alleviate the concerns of many in Parliament, particularly with regard to the morality and permissibility of human embryo research.[16] The VLA implemented its own code of practice and issued licences to doctors and scientists wishing to use IVF. Johnson describes the VLA as 'an attempt by the medical and scientific communities to regulate their own practice'.[17] However, the VLA itself soon felt that a statutory body and legislation regulating the sphere of IVF was

---

[10] Ibid 76.

[11] Ibid 76 [13.6].

[12] Ibid 66 [11.22].

[13] *Human Fertilisation and Embryology Act 1990* (UK) s 3(3)(a).

[14] The formation of the primitive streak is a major developmental milestone in a human embryo. It is a very early precursor to the central nervous system and is also the point beyond which an embryo can no longer divide and become twins (Department of Health and Social Security (UK), Warnock Report, 59).

[15] See, eg, United Kingdom, *Parliamentary Debates*, House of Lords, 13 September 2018, vol 792, col 2428 (Baroness Deech); Harris J, 'It's Time to Extend the 14-day Limit for Embryo Research', *The Guardian* (Online), 7 May 2016 < https://www.theguardian.com/commentisfree/2016/may/06/extend-14-day-limit-embryo-research > .

[16] Gunning J, English V (1993) Human In Vitro Fertilization: A Case Study in the Regulation of Medical Innovation (Dartmouth) 42. See also United Kingdom, *Parliamentary Debates*, House of Lords, 31 October 1984, vol 456, col 535 (The Marquess of Reading).

[17] Johnson MH (1998) 'Should the Use of Assisted Reproductive Techniques be Deregulated?' Human Reproduction 13: 1769, 1771.

required. Initially assuming it would only operate for 2 years, by which time legislation would be implemented and a statutory body would take over its functions, the VLA in fact regulated ART for six years.[18]

In 1991, the HFEA took over from the VLA. As originally conceived by the Warnock Committee, the HFEA has an advisory function and an executive function.[19] In its advisory capacity, the HFEA provides advice to the Secretary of State, clinics, patients and the public.[20] The HFEA's executive function centres primarily on the grant of three kinds of licences: treatment licences, storage licences and research licences.[21] Treatment licences authorise clinics to bring about the creation of an embryo in vitro, to test embryos and gametes and to place an embryo in a woman.[22] It is this kind of licence that will be the focus of this chapter. Storage licences authorise clinics to store embryos and gametes for a period of up to five years.[23] Research licences authorise research to be carried out on human embryos, as well as allowing human embryos to be created in vitro. [24]

As Deech and Smajdor note, an 'independent, central and transparent authority which grants licences, monitors research and imposes sanctions backed by criminal penalties, is a great asset'.[25] The importance of the HFEA was affirmed in 2012 and again in 2013, when it survived both a government proposal to transfer its functions to the Care Quality Commission[26] and then a proposed merger with the Human Tissue Authority.[27] On each occasion, the benefits of having an independent regulator for the fertility sector were noted.[28]

The establishment of the HFEA and the licensing system that it oversees were the key features of the *Human Fertilisation and Embryology Act 1990* (UK). In addition to establishing the HFEA, the 1990 Act set out a comprehensive regulatory

---

[18] Gunning J, English V (1993) Human In Vitro Fertilization: A Case Study in the Regulation of Medical Innovation (Dartmouth) 47.

[19] Department of Health and Social Security (UK), *Report of the Committee of Inquiry into Human Fertilisation and Embryology* (Cmd 9314, 1984) iv ('Warnock Report'), 76.

[20] *Human Fertilisation and Embryology Act 1990* (UK) ss 8(a)–(c).

[21] Ibid s 11(1).

[22] Ibid sch 2, para 1(1)(a)–(e).

[23] Ibid sch 2, para 2.

[24] Ibid sch 2, para 3(1).

[25] Deech R, Smajdor A, *From IVF to Immortality* (Oxford University Press, 2007) 6.

[26] Department of Health (UK), *Consultation on Proposals to Transfer Functions from the Human Fertilisation and Embryology Authority and the Human Tissue Authority* (2012).

[27] McCracken J, *Review of the Human Fertilisation & Embryology Authority and the Human Tissue Authority* (2013).

[28] Department of Health (UK), *Government Response to the Consultation on Proposals to Transfer Functions from the Human Fertilisation and Embryology Authority and the Human Tissue Authority* (2013) 7; Department of Health (UK), *Response to the Review of the HFEA and HTA: Government Response to the Independent Review of the Human Fertilisation and Embryology Authority and the Human Tissue Authority by Justin McCracken* (2013) 11.

framework for ART, including provisions covering parentage,[29] record keeping[30] and consent.[31] As part of this framework, the Act delineated clear limits on what were considered acceptable practices, with some activities being permitted only pursuant to a licence and others not permitted at all. For example, the creation,[32] keeping alive, and use of embryos,[33] were practices that were only permitted pursuant to a licence. Transferring a non-human embryo to a woman,[34] or placing gametes other than human gametes into a woman, were procedures that were completely prohibited.[35] Other prohibited practices included allowing an in vitro embryo to develop beyond fourteen days, or using an embryo after this time; placing an embryo in an animal; and keeping or using embryos in circumstances that contravene regulations.[36] The 1990 Act also amended the *Abortion Act 1967* (UK)[37] and the *Surrogacy Arrangements Act 1985* (UK).[38]

3.2.2 Reform

In 2008, the *Human Fertilisation and Embryology Act 1990* was amended by the *Human Fertilisation and Embryology Act 2008* (UK). As Caroline Flint observed during the public consultation stage of the reform process, 'it was never expected that the Act would remain forever unchanged in this area of fast moving science'.[39] As the 2008 Act was an amending Act, it attracted criticism from those who felt that the reform process would have been a good opportunity to completely rethink the regulatory framework for ART, rather than simply focusing on what needed to be changed.[40] The 2008 Act made a number of amendments to the original legislation

[29] *Human Fertilisation and Embryology Act 1990* (UK) ss 27–30.

[30] Ibid s 31.

[31] Ibid sch 3.

[32] Ibid s 3(1)(a).

[33] Ibid s 3(1)(b).

[34] Ibid s 3(2)(a).

[35] Ibid s 3(2)(b).

[36] Ibid ss 3(3)(a)–(c).

[37] Ibid s 37.

[38] Ibid s 36.

[39] Flint C, 'Foreword' in Department of Health (UK), R*eview of the Human Fertilisation and Embryology Act: A Public Consultation* (2005) 3. At the time, Caroline Flint was the Parliamentary Under Secretary of State for Public Health.

[40] McCandless J, Sheldon S (2010) The Human Fertilisation and Embryology Act (2008) and the Tenacity of the Sexual Family Form. The Modern Law Review 73: 175, 180; Fox M (2009) The Human Fertilisation and Embryology Act 2008: Tinkering at the Margins. Feminist Legal Studies 17(3): 333, 334.

to enable the law to keep pace with scientific developments[41] and changing societal norms. These changes included specific provisions relating to preimplantation genetic testing (PGT) and HLA typing,[42] amendment of the controversial welfare provision that governs who is eligible for treatment,[43] significant changes to the determination of parenthood[44] and provisions in respect of the creation of human admixed embryos.[45] As the provisions dealing with PGT, HLA typing and the welfare provision are discussed in the following sections, this section takes a closer look at the provisions relating to recognition of parenthood.[46]

The most significant changes to the parenthood provisions can be found in sections 42–44 of the 2008 Act. These provisions permit two women to be recognised as the legal parents of a child, from the moment of birth.[47] The Act states that the woman carrying the child is to be treated as the mother of that child.[48] McCandless and Sheldon note that at no point during the reform process was the idea that motherhood may not only be grounded in gestation given any serious consideration.[49] If the mother is married or in a civil partnership[50] at the time of receiving treatment, her partner will be recognised as the other parent of the child.[51] In this way, the Act brings parenthood for married same sex couples directly into line with married heterosexual couples.[52] Similarly, the parenthood provisions for unmarried same sex couples mirror those for unmarried heterosexual couples.[53] However, to satisfy the

---

[41] For example, the 2008 Act contains references to mitochondrial donation, and following the coming into force of the *Human Fertilisation and Embryology (Mitochondrial Donation) Regulations 2015* (UK) SI 2015/572, permits the use of eggs and embryos that contain donated mitochondria, 'in order to prevent the transmission of serious mitochondrial disease'. *Human Fertilisation and Embryology Act 2008* (UK) s 3ZA(5). These provisions made the UK the first country in the world to permit mitochondrial donation as part of IVF.

[42] *Human Fertilisation and Embryology Act 2008* (UK) sch 2, para 1ZA.

[43] Ibid s 14(2)(b).

[44] Ibid ss 33–47.

[45] Ibid s 4A.

[46] Although the provisions regarding the creation of human admixed embryos attracted significant media attention, their relevance to research, as opposed to treatment, puts them beyond the scope of this chapter. See, eg, Prince R, 'MPs Vote Against Human-Animal Hybrid Embryo Ban', *The Telegraph* (online), 19 May 2008 < http://www.telegraph.co.uk/news/politics/1988846/MPs-vote-against-human-animal-hybrid-embryo-ban.html >; BBC News, *MPs back Hybrid Embryo Research* (19 May 2008) http://news.bbc.co.uk/2/hi/uk_news/politics/7407589.stm. Accessed 17 December 2024.

[47] *Human Fertilisation and Embryology Act 2008* (UK) s 42.

[48] Ibid s 33(1). This provision mirrors s 27 of the *Human Fertilisation and Embryology Act 1990* (UK).

[49] McCandless J, Sheldon S (2010) The Human Fertilisation and Embryology Act (2008) and the Tenacity of the Sexual Family Form. The Modern Law Review 73: 175, 183.

[50] As it is possible to convert a civil partnership into a marriage in England and Wales, the term 'married' will be used to refer to both married and civilly partnered same sex couples.

[51] *Human Fertilisation and Embryology Act 2008* (UK) s 42(1).

[52] Ibid s 35(1).

[53] Ibid s 43 and s 36 respectively.

'female parenthood conditions', the Act requires unmarried couples to have received treatment at a licensed clinic; a condition not imposed on married same sex couples.[54] The different ways of treating married and unmarried couples has left the 2008 Act open to the criticism that it maintains the same 'hierarchical structure' as the 1990 Act.[55] Thus while recognition of same sex relationships is a step in the right direction, the placement of marriage at the top of the pyramid is perhaps not reflective of the lived experiences of many couples. Further, as McCandless and Sheldon point out, 'the ongoing distinction between married and unmarried [couples] is interesting in the context of widespread erosion of the relevance of this distinction elsewhere in family law'.[56]

The parenthood provisions of the 2008 Act have also attracted criticism for adopting a 'two parent model'.[57] Objections to this model have centred on the Act not permitting a child to have three legal parents, thus continuing to reflect a 'heteronormative family form'.[58] Furthermore, criticism has been directed at the fact the 2008 Act does not permit registration of two mothers; rather, a child can have only one mother[59] and a 'female parent'. The practical effect of this provision means the parent carrying the baby is considered the mother and her partner, the 'female parent'. McCandless and Sheldon argue that this model encompasses what they refer to as 'parental dimorphism', meaning that parents are viewed as occupying 'complementary yet different legal roles'.[60] Thus it may be argued that this is yet another example of the 'heteronormative family form' ideal underpinning the Act.

3.3 Eligibility for Treatment

Instead of criteria determining eligibility for treatment, as can be found in many Australian states, the *Human Fertilisation and Embryology Act 1990* (UK) opted to restrict access to ART through a welfare provision purporting to be concerned with the future child's welfare. Section 13(5) of the 1990 Act prohibited ART clinics from treating a woman unless they had taken into account 'the welfare of any child who may been born as a result of the treatment (including the need of that child for a father)'. Interestingly, the wisdom of including such a provision in the legislation

---

[54] Ibid s 43(a), s 44(1).

[55] McCandless J, Sheldon S (2010) The Human Fertilisation and Embryology Act (2008) and the Tenacity of the Sexual Family Form. The Modern Law Review 73: 175, 188.

[56] Ibid 190.

[57] Ibid 190–193.

[58] Fox M (2009) The Human Fertilisation and Embryology Act 2008: Tinkering at the Margins. 17(3) *Feminist Legal Studies* 333, 337. See also Zanghellini A (2010) Lesbian and Gay Parents and Reproductive Technologies: The 2008 Australian and UK Reforms 18(3) *Feminist Legal Studies* 227, 228.

[59] *Human Fertilisation and Embryology Act 2008* (UK) s 33(1).

[60] McCandless J, Sheldon S (2010) The Human Fertilisation and Embryology Act (2008) and the Tenacity of the Sexual Family Form. The Modern Law Review 73: 175, 193.

received little debate during the passage of the 1990 Act and as Jackson asserts, 'it was simply assumed to be self-evidently true that their future children's welfare ought to be taken into account before a couple is offered assistance with conception'.[61] This provision has, however, proved controversial for two reasons; first, it required (and still requires) clinics to make a determination about the welfare of a child who has not yet come into existence. Second, the provision discriminated against lesbian couples and single women.

Eighteen years later, section 13(5) was updated by the 2008 Act, which kept the provision virtually unchanged except for substituting the words 'need of that child for a father' with a requirement that clinics take into account a child's need for 'supportive parenting'.[62] While it has been suggested that this change in terminology has little impact in practice,[63] the removal of explicit reference to 'a father' generated a great deal of debate both in the media[64] and in Parliament. Arguments in Parliament varied from assertions that 'it would be wrong of the state to facilitate the deliberate creation of children with the intention that they should be denied the chance of ever having a father for the duration of their childhood',[65] to arguments based on the detrimental effect on a child of not having both a father and a mother.[66] McCandless and Sheldon, however, note that arguments based on research conducted into children from 'broken homes' or unplanned pregnancies should not be conflated with cases where a woman or lesbian couple have deliberately created a child who will not have a father.[67] This is due to many cases of unplanned pregnancy being accompanied by poverty or financial hardship, whereas these factors are not usually present in the case of a single woman who has made the informed decision to bring up a child on

[61] Jackson E (2002) Conception and the Irrelevance of the Welfare Principle 65 *The Modern Law Review* 176, 176.

[62] *Human Fertilisation and Embryology Act 2008* (UK) s 14(2)(b).

[63] Sheldon S, Lee E, Macvarish J, (2015) 'Supportive Parenting', Responsibility and Regulation: The Welfare Assessment under the Reformed Human Fertilisation and Embryology Act (1990). The Modern Law Review 78: 461, 462.

[64] See eg, Sarah Hall, 'IVF Hurdle for Single Women and Lesbians to be Overthrown', *The Guardian* (online) 13 July 2006 < https://www.theguardian.com/uk/2006/jul/13/health.familyandrelationships > ; Nicholas Watt, 'MPs Reject 'Need for Father' in IVF Bill', *The Guardian* (online) 21 May 2008 < https://www.theguardian.com/politics/2008/may/20/health.health > ; Graeme Wilson, 'No Father Required', *Daily Mail* (online) 15 August 2005 < http://www.dailymail.co.uk/health/article-359287/No-father-required.html > ; Ben Russell and Nigel Morris, 'Fertility Laws Drop Need to consider Role of Father', *Independent* (online) 21 May 2008 < http://www.independent.co.uk/news/uk/politics/fertility-laws-drop-need-to-consider-role-of-father-831544.html > .

[65] United Kingdom, *Parliamentary Debates*, House of Commons, 3 June 2008, vol 476, col 227 (Gary Streeter).

[66] Conservative politician Iain Duncan Smith, for example, cited research indicating that children without fathers 'are more likely to fail at school… succumb to drug addiction' and have serious alcohol problems. United Kingdom, *Parliamentary Debates*, House of Commons, 20 May 2008, vol 476, col 167 (Iain Duncan Smith).

[67] McCandless J, Sheldon S (2010) '"No Father Required"? The Welfare Assessment in the Human Fertilisation and Embryology Act 2008' 18 *Feminist Legal Studies* 201, 211.

her own.[68] Similarly, research indicates that children born to lesbian couples do no worse across a range of performance indicators than children born to heterosexual couples.[69]

Guidance for clinics on how to apply the updated welfare provision can be found in the 9th edition of the HFEA's Code of Practice. The Code of Practice requires clinics to conduct their welfare assessments in a non-discriminatory manner and not discriminate against prospective parents on the basis of marital status, gender, race, disability, sexual orientation, religious belief or age.[70] It also lists a number of factors that clinics should take into account when carrying out their assessment, including:

> (a) past or current circumstances that may lead to any child… experiencing serious physical or psychological harm or neglect, for example:
>
> (i) previous convictions relating to harming children
>
> (ii) child protection measures taken regarding existing children, or
>
> (iii) violence or serious discord in the family environment
>
> (b) past or current circumstances that are likely to lead to an inability to care throughout childhood for any child who may be born, or that are already seriously impairing the care of any existing child of the family, for example:
>
> (i) mental or physical conditions
>
> (ii) drug or alcohol abuse
>
> (iii) medical history, where the medical history indicates that any child who may be born is likely to suffer from a serious medical condition, or
>
> (iv) circumstances that the centre considers likely to cause serious harm to any child [who may be born or any existing child].[71]

Further, the Code of Practice defines what is meant by supportive parenting. It provides that.

> '[s]upportive parenting is a commitment to the health, well being and development of the child. It is presumed that all prospective parents will be supportive parents, in the absence of any reasonable cause for concern that any child who may be born, or any other child, may be at risk of significant harm or neglect'.[72]

The Code of Practice also permits clinics to look at the broader family context in which the child will be raised, in the event that they have concerns about the existence of supportive parenting arrangements. Furthermore, clinics are advised to seek patients' consent in order to make further enquiries, should any concerns arise as a result of the information given by a patient or their partner.[73] In the event that a

---

[68] Ibid 210.

[69] See eg, MacCallum F, Golombok S (2004), Children Raised in Fatherless Families From Infancy: a follow-up of children of lesbian and single heterosexual mothers at early adolescence 45 *Journal of Child Psychology and Psychiatry* 1407.

[70] Human Fertilisation and Embryology Authority, *Code of Practice* (9th ed, revised 2023) [8.6].

[71] Ibid [8.14].

[72] Ibid [8.15].

[73] Ibid [8.16].

patient refuses to consent, the Code of Practice specifies that such a refusal 'should not, in itself, be grounds for denying treatment but the centre should take this into account'.[74] The Code's emphasis on non-discrimination, the wide construction given to the phrase 'supportive parenting' and the procedures surrounding the collection of further information, has led some academics to refer to the Code as containing a 'presumption to treat'.[75]

The current version of the welfare provision goes some way to addressing the deficiencies of its predecessor. However, while the 2008 Act removes what was seen by many single women and lesbian couples as a barrier to conception, the welfare provision still faces the criticism that both of these groups and the clinically infertile are still required to prove their parenting capacity when other prospective parents are not.

3.4 Selection Against Disability

This section considers the regulation of PGT-M for the purposes of selection against disability in the UK. It first considers the original provisions of the 1990 Act, before considering the current framework and guidance following the 2008 amendments.

The *Human Fertilisation and Embryology Act 1990* (UK) was drafted at a time when PGT was in its infancy. As the first successful birth following IVF and PGT did not occur until 1990,[76] the Act only contained limited references to embryo testing. The 1990 Act permitted PGT under licence pursuant to schedule 2, paragraph 1(1)(d), as PGT would be considered a practice 'designed to secure that embryos are in a suitable condition to be placed in a woman or to determine whether embryos are suitable for that purpose'. In 2007, the *Human Fertilisation and Embryology (Quality and Safety) Regulations 2007* (UK) amended schedule 2, paragraph 1 of the 1990 Act to specifically include the testing of embryos within the list of activities that may be authorized under a treatment licence.[77]

Schedule 2 of the 1990 Act was further amended by the *Human Fertilisation and Embryology Act 2008* (UK), which inserted more detailed embryo testing provisions into the legislation. As the scope of PGT had substantially increased since the original Act, the provisions contained in the 2008 Act are predominantly prohibitive in nature. For example, the 2008 Act inserted paragraph 1ZA into the 1990 Act. This paragraph prohibits PGT except where it is necessary to establish whether the embryo 'has a gene, chromosome or mitochondrion abnormality that may affect its capacity to

---

[74] Ibid.

[75] Sheldon S, Lee E, Macvarish J, (2015) 'Supportive Parenting', Responsibility and Regulation: The Welfare Assessment under the Reformed Human Fertilisation and Embryology Act (1990). The Modern Law Review 78: 461, 465.

[76] S G Hillier, 'Training and Accreditation of IVF Embryologists' in A A Templeton and J O Drife (eds), *Infertility* (Springer-Verlag, 1992) 363, 365. See also AH Handyside et al., 'Biopsy of Human Preimplantation Embryos and Sexing by DNA Amplification' (1989) 333(8634) *The Lancet* 347.

[77] *Human Fertilisation and Embryology (Quality and Safety) Regulations 2007* (UK) reg 29(a)(i).

result in a live birth'[78] or to establish whether an embryo that is at a 'particular risk' of developing an abnormality has that abnormality.[79] Further, the Act specifies that for testing to be permitted on this second ground, there must exist 'a significant risk that a person with the abnormality will have or develop a serious physical or mental disability, a serious illness or any other serious medical condition'.[80] In Parliament, it was suggested that the word 'will' in paragraph 1ZA(2) be replaced with the word 'may'.[81] The result of this amendment would have been that for testing to be permitted it only needed to be shown that a 'significant risk' of disability 'may' result following an abnormality in an embryo. On the face of it, the purpose of the amendment appears to be to allow lower penetrance conditions, however, the inclusion of the term 'significant risk' suggests that the amendment would not have altered the level of penetrance required. Further, it was argued that the amendment would have added an undesirable level of uncertainty.[82] As it is unlikely that the amendment would have made any difference to the operation of the legislation, it is unsurprising that it did not make it into the Act.

In addition to the legislation, further guidance on the uses of PGT may be found in the HFEA's Code of Practice. The Code of Practice re-states the legislative position that PGT may only be undertaken to prevent the development of a 'serious disability, illness or medical condition' in the embryo, or to prevent a sex-linked condition.[83] With respect to PGT for the purposes of selection against disability, the guidance provided in the Code of Practice suggests that clinics 'should consider the circumstances of those seeking treatment rather than the particular heritable condition'.[84] While no further guidance is provided as to what circumstances would be relevant to such a determination, relevant circumstances may include the prospective parents' financial circumstances, whether they already have a child/children, and whether any of those existing children have the condition for which PGT is being requested. Therefore, one set of prospective parents may be granted PGT for a specific condition, whilst another set are denied PGT for the same condition. The Code of Practice thus gives clinics significant discretion in determining who should be allowed access to PGT, and it is questionable whether clinic staff are in the best position to pass judgement on the circumstances of an individual or couple seeking treatment.

The Code of Practice does, however, require clinics to discuss the seriousness of the condition for which PGT is sought with the prospective parents prior coming to a decision as to whether to offer PGT.[85] Interestingly, neither the legislation nor the

---

[78] *Human Fertilisation and Embryology Act* 1990 (UK) sch 2 para 1ZA(1)(a).

[79] Ibid para 1ZA(1)(b).

[80] Ibid para 1ZA(2).

[81] United Kingdom, *Parliamentary Debates*, House of Lords, 4 December 2007, vol 696, col 1647 (Baroness Royall).

[82] Ibid.

[83] Human Fertilisation and Embryology Authority, *Code of Practice* (9th ed, revised 2023) box 10A.

[84] Ibid [10.5].

[85] Ibid [10.6].

Code of Practice defines what is meant by a serious condition. This could, in theory, promote a flexible approach to decision-making and allow a couple or individual seeking treatment a greater say in what they would consider a serious condition. There is, however, an additional level of scrutiny that needs to be overcome with regard to new conditions. The HFEA only permits clinics to test for a novel condition if it meets the criteria set out in the Act.[86] This decision is made by the HFEA's statutory approvals committee (SAC). Thus, the decision as to the seriousness of a condition is taken out of both the clinic's and the prospective parents' hands. This extra level of scrutiny was criticized by the House of Commons Science and Technology Committee prior to the 2008 Act coming into force. In its report on ART, *Human Reproductive Technologies and the Law*, the Committee argued that there was 'no reason why a regulator should seek to determine which disorders can be screened out using PGD'.[87] Unfortunately, this issue did not receive any scrutiny during the passage of the legislation through Parliament.

The HFEA's approval process for conditions that have not previously been licensed has been a barrier for some prospective parents. In 2014, the SAC declined approval for autism spectrum disorder (ASD) to be added to the list of approved conditions.[88] The request came from a clinic that wished to use PGT to select against ASD in families where at least two sons have been diagnosed with severe ASD.[89] As there is no genetic test for ASD, the clinic wanted to use sex selection to select a female embryo, as ASD is four times more likely in males than females.[90] As ASD is not sex linked in the same way as, for example, haemophilia, there was no guarantee that a female child would not have ASD; just a significantly lower chance. As a result, the SAC did not feel that using PGT for this purpose met the statutory criteria in paragraph 1ZA(1)(c).[91] Further, the SAC found a lack of compelling evidence to prove that in a particular family with two children affected by autism, that a third child, even if that child was a girl, would not be similarly affected. The SAC therefore concluded that there did not exist 'a particular risk' that would be prevented by the use of PGT.[92]

---

[86] HFEA, PGT-M Conditions <https://www.hfea.gov.uk/pgt-m-conditions/>.

[87] Science and Technology Committee, *Human Reproductive Technologies and the Law*, House of Commons Paper No 7-I, Session 2004–05 (2005) [124]. This report informed the review of the law that culminated in the *Human Fertilisation and Embryology Act 2008* (UK). The terms of reference for the Science and Technology Committee's inquiry included, to consider how reproductive freedom should be balanced with regulation and legislation; to consider what Parliament's role in the regulation of ART; and to consider the 'foundation, adequacy and appropriateness of the ethical framework for legislation' on ART. Further, the committee was required to consider the provisions of the 1990 Act in relation other national and international laws that regulate the sphere of health law: 4.

[88] Statutory Approvals Committee, HFEA 'Minutes' (30 January 2014) [20].

[89] Ibid [3].

[90] For a contrary view, see Constantino JN, Charman T (2012), Gender Bias, Female Resilience, and the Sex Ratio in Autism. 51 *Journal of the American Academy of Child and Adolescent Psychiatry* 756.

[91] Statutory Approvals Committee, HFEA 'Minutes' (30 January 2014) [18].

[92] Ibid [16].

Once a condition has been approved by the HFEA as having satisfied the statutory requirements, the Code of Practice requires clinics to make a decision on whether it is 'appropriate' to use PGT in a particular case. This creates an additional hurdle for patients over and above those contained in the legislation. The Code of Practice also sets out a list of criteria which clinics should consider when deciding on the appropriateness of PGT in a particular case:

(a) the views of the people seeking treatment in relation to the condition to be avoided, including their previous reproductive experience
(b) the likely degree of suffering associated with the condition
(c) the availability of effective therapy, now and in the future
(d) the speed of degeneration in progressive disorders
(e) the extent of any intellectual impairment
(f) the social support available, and
(g) the family circumstances of the people seeking treatment.[93]

It could be argued that this kind of comprehensive guidance should be welcomed as it promotes consistency in the decision-making process between clinics. However, the Science and Technology Committee had 'concerns about the criteria imposed by the HFEA'.[94] The exact nature of these concerns is not detailed in the committee's report, but it may be the existence of the criteria themselves that is deemed objectionable, rather than any specific factor on the list. This is backed up by recommendation 60 of the Science and Technology Committee's report, in which they 'recognise that the HFEA has legal jurisdiction but this does not mean that it has a duty to regulate [PGT] beyond ensuring that it is performed to the highest standards within statutory boundaries'.[95]

The criteria listed in the Code of Practice may also be problematic for another reason: rather than promoting consistency in decision-making, they may lead to inconsistent outcomes. For example, it is unclear how 'the family circumstances of the people seeking treatment'[96] are relevant to whether a couple or individual is eligible for PGT. If a couple has existing children, does this increase their suitability and thus make the couple eligible for PGT? Conversely, would a couple without existing children be considered as more able to cope with a child with the same condition and thus not eligible for PGT? Nevertheless, it may be argued that criteria which take into account family circumstances or social support are preferable to a list of criteria limited to the medicalised elements of a disability or condition. However, without additional guidance regarding the interpretation and balancing of these factors, there is a risk that any potential usefulness will be undermined.

---

[93] Human Fertilisation and Embryology Authority, *Code of Practice* (9th ed, revised 2023) [10.9].

[94] Science and Technology Committee, *Human Reproductive Technologies and the Law, Human Reproductive Technologies and the Law*, House of Commons Paper No 7-I, Session 2004–05 (2005). It should be noted that these criteria are the same as those in the sixth edition of the Code of Practice – the most recent edition at the time of the committee's report.

[95] Ibid [244].

[96] HFEA, *Code of Practice* (9th ed, revised 2023) [10.9(g)].

The regulation of PGT for the purposes of selection against disability needs to be reviewed. In particular, the data from clinics regarding which conditions are deemed appropriate for PGT, and the circumstances of the couple or individual to whom this is granted, requires analysis. This analysis should focus on the decision-making process that occurs within clinics, in order to highlight any inconsistencies between different units. Further, as noted above, the ultimate decision as to whether to offer PGT lies with clinics; this is unacceptable when it is the lives of the prospective parents that will be affected by the birth of a child with a disability. Decisions regarding whether or not to use PGT need to be made by patients, with guidance from clinic staff, and within broad parameters set by Parliament.

Another way in which the regulatory framework could be improved is by removing the word 'serious', in the context of 'serious condition' or 'serious disability', from the legislation and supporting framework. Ascertaining what counts as a serious condition in individual cases is problematic for clinicians and not a question capable of a definitive answer.[97] For example, while breast cancer may be considered to be a serious illness, one clinician giving evidence to the Science and Technology Committee suggested that tests for the BRCA1 and BRCA2 genes that are linked to susceptibility for breast cancer, would fall outside the definition of serious.[98] Further, if PGT is acceptable for serious conditions, why is it not acceptable for all conditions that cause any kind of suffering?[99]

3.5 HLA Typing

This section looks at the regulation of HLA typing in the United Kingdom. It first examines the regulatory framework for HLA typing under the 1990 Act, then charts its evolution until the *Human Fertilisation and Embryology Act 2008* (UK).

At the time the 1990 Act was enacted, HLA typing had not as yet been developed and thus was not a target for regulation. In 2001, a year after the world's first saviour sibling was born in the United States, the HFEA received its first application from a clinic wishing to use tissue typing in conjunction with PGT. The application was made on behalf of the parents of Zain Hashmi, a 4-year-old boy with beta thalassaemia. His parents wished to conceive a saviour sibling for Zain, who would be free from beta thalassaemia and also able to donate umbilical cord stem cells to Zain. The HFEA granted a treatment licence to the clinic on the basis that because PGT was required to prevent the future child from inheriting beta thalassaemia, HLA typing to determine tissue compatibility could also be justified.[100]

---

[97] Science and Technology Committee, *Human Reproductive Technologies and the Law*, House of Commons Paper No 7-II, Session 2004–05 (2005) Ev 10 (Richard Kennedy).

[98] Ibid.

[99] Ibid Ev 80 (Desmond Turner).

[100] HFEA, 'HFEA to Allow Tissue Typing In Conjunction with Preimplantation Genetic Diagnosis' (Press Release, 13 December 2001) < http://www.hfea.gov.uk/961.html > .

Shortly after the grant of this licence, Josephine Quintavalle, on behalf of Comment on Reproductive Ethics (CORE), applied for judicial review of the HFEA's decision to grant the licence, on the grounds that the HFEA had no power to authorise tissue typing.[101] Quintavalle argued that HLA typing fell outside the HFEA's remit as pursuant to schedule 2 paragraph 1(3), a treatment licence can only authorise activities that are 'necessary or desirable' for the provision of treatment services.[102] Defining treatment services as services 'for the purpose of assisting women to carry children', Quintavalle claimed that tissue typing did not assist a woman to become pregnant with or carry a child, but was instead for the purposes of curing an existing child.[103] The HFEA, on the other hand, argued that tissue typing did not require a licence because it is not a procedure carried out on an embryo, but instead it is carried out on cells already extracted from an embryo[104] for the purposes of PGT. Therefore, it fell outside the scope of the prohibitions in s 3(1)(b) of the 1990 Act.

At first instance, Maurice Kay J rejected the HFEA's argument on the basis that it was 'inconceivable' that Parliament would have intended to leave tissue typing beyond the scope of the Act.[105] Therefore, the distinction made by the HFEA between cells extracted from an embryo and an embryo itself was dismissed as 'artificial, unattractive and unnecessary'.[106] As tissue typing was deemed to fall within the scope of the Act, the next question was whether it could be authorised under the Act. Maurice Kay J's response was that as carrying a child was not affected by tissue typing, it fell outside the meaning of treatment services in s 2(1) of the 1990 Act.[107] The case proceeded all the way to the House of Lords where it was held that PGT and tissue typing could lawfully be authorised by the HFEA as activities to determine the suitability of an embryo within the meaning of schedule 2, paragraph 1(1)(d) of the 1990 Act.[108] Further, their Lordships noted that Parliament had not intended to confine the HFEA's powers to the determination of unsuitability on the grounds of genetic defects. Parliament's intention was to leave it to the HFEA to decide whether activities such as tissue typing could be permitted.

Brownsword suggests that the House of Lords decision in this case was based on a 'quite extraordinary reading of the legislation'.[109] In his judgment, Lord Hoffman considered the word 'suitable' in schedule 2 paragraph 1(1)(d) to be an adjective that is context dependent. Quoting Lord Scott of Foscote, Lord Hoffman noted that 'a

---

[101] *R (on the application of Josephine Quintavalle on behalf of Comment on Reproductive Ethics) v Human Fertilisation & Embryology Authority* [2002] EWHC 2785.

[102] Ibid [6].

[103] Ibid.

[104] Ibid.

[105] Ibid [12].

[106] Ibid.

[107] Ibid [17].

[108] *Quintavalle (on behalf of Comment on Reproductive Ethics) v Human Fertilisation and Embryology Authority* [2005] UKHL 28.

[109] Brownsword R (2005) Happy Families, Consenting Couples, and Children with Dignity: Sex Selection and Saviour Siblings. 17 *Child and Family Law Quarterly* 435, 462.

suitable hat for Royal Ascot is very different from a suitable hat for the Banbury cattle market'.[110] Therefore, what needed to be determined according to Lord Hoffman was the extent to which the concept of suitability took into account the needs of the mother. Brownsword suggests that 'the key to cracking the legislative code' was in fact the phrase 'that purpose' in paragraph 1(1)(d) of the 1990 Act.[111] The purpose in question could refer to the purpose of the particular woman, in this case Mrs Hashmi, or it could refer to a more generic purpose of assisting women to carry a child. As noted above, the House of Lords chose what Brownsword refers to as the 'woman specific' interpretation.[112] Their Lordships' judgment may also be referred to as a 'pragmatic attempt to interpret the terms of the 1990 Act in such a way as to render it workable in the light of evolving science'.[113] The pragmatism shown by the House of Lords is also desirable because, as Sally Sheldon notes, it avoids a situation 'whereby it would be illegal to select an embryo for tissue compatibility yet perfectly lawful to allow Mrs Hashmi to conceive a series of pregnancies and terminate those where testing revealed the developing foetus to be an incompatible tissue donor'.[114]

Shortly after the HFEA granted the Hashmis permission for tissue typing, they received a similar request from the Whitakers. Charlie Whitaker suffered from diamond blackfan anaemia (DBA), a rare condition that causes the bone marrow to produce insufficient red blood cells and in some cases, no red blood cells. The Whitakers wished to conceive a saviour sibling for Charlie. The Whitaker case contained one significant difference to that of the Hashmis. Whereas Zain's condition was inherited, Charlie's was not. This meant that a child conceived naturally by the Whitakers was no more likely to suffer from DBA than any of the general population. As a result, PGT was unnecessary and only HLA typing was required. Therefore, embryo testing was solely for the benefit of Charlie, not the future child.

This distinction between HLA typing following PGT, and HLA typing on its own was an important one for the HFEA. In granting a licence to the Hashmis, Ruth Deech, the Chair of the HFEA at the time of the decision, stated that, 'where PGD is already being undertaken we can see how the use of tissue typing to save the life of a sibling could be justified'.[115] As PGT was not required for the embryos created by the Whitakers, the HFEA refused to grant a licence for HLA typing.[116] In coming

---

[110] *Quintavalle (on behalf of Comment on Reproductive Ethics) v Human Fertilisation and Embryology Authority* [2005] UKHL 28 [14].

[111] Brownsword R (2005) Happy Families, Consenting Couples, and Children with Dignity: Sex Selection and Saviour Siblings. 17 *Child and Family Law Quarterly* 435, 462.

[112] Ibid.

[113] Sheldon S (2005), Saviour Siblings and the Discretionary Power of the HFEA. 13 *Medical Law Review* 403, 411.

[114] Ibid 410.

[115] Human Fertilisation and Embryology Authority, 'HFEA to Allow Tissue Typing In Conjunction with Preimplantation Genetic Diagnosis' (Press Release, 13 December 2001) < http://www.hfea.gov.uk/961.html > .

[116] Human Fertilisation and Embryology Authority, 'HFEA confirms that HLA tissue typing may only take place when PGD is required to avoid a serious genetic disorder' (Press Release, 1 August 2002) < http://www.hfea.gov.uk/935.html > .

to this decision, the HFEA ignored the recommendation made by its own ethics committee, which advised that tissue typing 'should be available where there is an existing sibling with a life-threatening but non-inherited condition'.[117] However, the HFEA decided that as the future child did not gain a benefit from the removal of a cell from the embryo for testing, 'the balance of potential harm and potential goods falls in a different place' than in cases where PGT is also required.[118] As a consequence, the Whitakers were unable to continue treatment in the UK and had to go to Chicago, where they conceived a saviour sibling for Charlie.

The HFEA's decision not to grant the Whitakers a licence was heavily criticised, with some academics referring to the distinction drawn between the Hashmi and Whitaker cases as 'unjustifiable and misguided'.[119] Sheldon and Wilkinson argue that if the HFEA's reasoning was based on a benefit to the embryo from PGT, that benefit must be existence rather than being free from a genetic disorder.[120] This is because PGT does not make an affected embryo healthy, but rather causes it to exist. Therefore, they assert that the embryos created by the Hashmis and those created by the Whitakers both stood to 'benefit' from the selection process because the testing would result in them being born, or at least having the opportunity to become a child, rather than not existing.[121] Sheldon and Wilkinson also note that the HFEA's reasoning gives rise to the further difficulty of whether existence is itself a benefit.[122]

In July 2004, the HFEA announced that it had extended its policy on tissue typing to cases where tissue typing was required without PGT.[123] Following a review of the risks to an embryo from biopsy, the HFEA concluded that

> the risk to the resulting child associated with embryo biopsy is not sufficient enough to warrant a policy which distinguishes between cases in which preimplantation tissue typing is used in combination with PG[T] for serious disease and where discovering tissue type is the sole treatment objective.[124]

Interestingly, the HFEA's report may also open the door to the conception of a saviour sibling for a child who, while currently healthy, may be in remission from a disease and there exists the possibility of a relapse.[125]

---

[117] Ethics Committee, HFEA, *Ethical Issues in the Creation and Selection of Preimplantation Embryos to Produce Tissue Donors* (22 November 2001) [3.14].

[118] Suzi Leather, quoted in 'Head–to–Head: Designer Babies' (19 June 2003) BBC News < http://news.bbc.co.uk/2/hi/health/3002858.stm > .

[119] Sheldon S, Wilkinson S (2004) Hashmi and Whitaker: An Unjustifiable and Misguided Distinction? 12 *Medical Law Review* 137. See also Gavaghan C (2003) Use of Preimplantation Genetic Diagnosis to Produce Tissue Donors: An Irreconcilable Dichotomy? 6 *Human Fertility* 23.

[120] Sheldon S and Wilkinson S, 'Hashmi and Whitaker: An Unjustifiable and Misguided Distinction?' (2004) 12 *Medical Law Review* 137, 156.

[121] Ibid 157.

[122] Ibid.

[123] Human Fertilisation and Embryology Authority, 'HFEA agrees to extend policy on tissue typing' (Press Release, 21 July 2004) < http://www.hfea.gov.uk/763.html > .

[124] HFEA, *Preimplantation Tissue Typing* (Report, July 2004) 5.

[125] Ibid 9.

The current law on HLA typing can be found in schedule 2 paragraph 1ZA(1)(d) of the *Human Fertilisation and Embryology Act 1990* (UK). This paragraph permits embryo testing for the purposes of establishing tissue type in cases where a sibling of the future child 'suffers from a serious medical condition which could be treated by umbilical cord blood stem cells, bone marrow or other tissue of any resulting child'.[126] The Act also specifies that 'other tissue' does not refer to whole organs.[127] Thus, HLA typing will not be available where the sick sibling suffers from renal failure and needs a kidney. In Parliament, it was noted that the term 'other tissue' was 'intended to capture regenerative tissue other than whole organs.'[128] Further guidance on the interpretation of this section is found in the HFEA's Code of Practice, which provides that in addition to considering each case individually,[129] clinics should take into account a number of factors affecting the sick child including:

(a) the degree of suffering associated with their condition
(b) the speed of degeneration in progressive disorders
(c) the extent of any intellectual impairment
(d) their prognosis, considering all treatment options available
(e) the availability of alternative sources of tissue for treating them, now and in the future, and
(f) the availability of effective therapy for them, now and in the future.[130]

The Code of Practice also requires clinics to consider the impact of tissue typing on the saviour sibling,[131] including any risks associated with embryo biopsy, long-term emotional or psychological consequences, the nature of the treatment (for example, whether it is invasive treatment such as bone marrow extraction) and 'any complications or predispositions associated with the tissue type selected'. Of these factors, determining the long-term emotional and psychological effects of HLA typing would be the most challenging for clinics, as it would be extremely difficult (if not impossible) to assess the future emotional reactions of a child that has yet to be born. This is also one of the most controversial aspects of HLA typing and formed the basis of an amendment in the House of Lords to delete paragraph 1ZA(1)(d) from the 2008 Act. Even though the amendment was defeated by a landslide, many of the risks detailed by supporters of the amendment will need to be carefully evaluated by clinics, including the vulnerable position of the savour sibling,[132] and the moral pressure they will face in the event of umbilical cord stem cells not curing the sick sibling and further donation being needed.[133]

---

[126] *Human Fertilisation and Embryology Act 1990* (UK) schedule 2, para 1ZA(1)(d).

[127] Ibid para 1ZA(4).

[128] United Kingdom, *Parliamentary Debates*, House of Lords, 15 January 2008, vol 696, col 1283. (Baroness Royall).

[129] Human Fertilisation and Embryology Authority, *Code of Practice* (9th ed, revised 2023) [10.24].

[130] Ibid [10.25].

[131] Ibid [10.26].

[132] United Kingdom, *Parliamentary Debates*, House of Lords, 15 January 2008, vol 696, col 1270. (Baroness O'Cathain).

[133] United Kingdom, *Parliamentary Debates*, House of Lords, 15 January 2008, vol 696, col 1274.

Furthermore, the Code of Practice instructs clinics to take into account the family circumstances of the couple or individual seeking treatment, including:

(a) their previous reproductive experience
(b) their views and the affected child's views of the condition
(c) the likelihood of a successful outcome, taking into account:

    (i) their reproductive circumstances (ie, the number of embryos likely to be available for testing in each treatment cycle, the number likely to be suitable for transfer, whether carrier embryos may be transferred, and the likely number of cycles)
    (ii) the likely outcome of treatment for the affected child

(d) the consequences of an unsuccessful outcome
(e) the demands of IVF/preimplantation testing treatment on them while caring for an affected child, and
(f) the extent of social support available.[134]

The Code of Practice also sets out a detailed list of information that should be provided to people seeking treatment, including the likelihood of a successful outcome for the sick child and the availability of counselling and social support.[135] Interestingly, the Code of Practice requires clinics to have procedures in place to enable them to follow-up with patients and their families, in particular, to provide the saviour siblings themselves with the opportunity to take part in 'long-term medical and psychosocial follow-up studies'.[136] There appears to be limited information pertaining to the results of these follow-up studies, which is unfortunate as they should play an important role in the regulation of HLA typing. The first-hand experiences of children born as saviour siblings would provide much-needed insight into the welfare of the child, which is the paramount concern of regulation in this sphere.

3.6 Non-Medical Sex Selection

This section examines the regulation of non-medical sex selection in the UK. Non-medical sex selection, also known as social sex selection, refers to the selection of embryos of a particular sex for reasons other than the avoidance of a disability. Non-medical sex selection is a kind of positive selection; whereas medical sex selection, like other kinds of selection against disability, is classed as negative selection.

The current law on sex selection can be found in schedule 2 paragraph 1ZA(1)(c) of the *Human Fertilisation and Embryology Act 1990* (UK), which provides that a treatment licence cannot authorise embryo testing for the purposes of establishing

(Baroness Williams).

[134] Human Fertilisation and Embryology Authority, *Code of Practice* (9th ed, revised 2023) [10.27].

[135] Ibid [10.28].

[136] Ibid [10.30].

the sex of an embryo, unless, 'there is a particular risk that any resulting child will have or develop' a serious 'gender related'[137] disability, illness or medical condition. This section implies that the use of PGT for non-medical sex selection is prohibited. Further, the Code of Practice explicitly provides that clinics should not 'for social reasons… select embryos of a particular sex'.[138]

Whether or not non-medical sex selection should be prohibited in the UK has been the subject of much debate. In 1993, the HFEA conducted a public consultation exercise to determine if sex selection using PGT should be permitted for medical and non-medical purposes.[139] The results of the consultation were that while PGT should be permitted to prevent a sex-linked medical condition, it should not be permitted for non-medical reasons.[140] Nine years later, the government launched a further public consultation into sex selection.[141] This time, the focus was very much on non-medical sex selection.

The 2002 consultation has been criticised for being 'biased towards certain outcomes', 'an attempt not to investigate but to influence public opinion'[142] and a 'rather meaningless' public relations exercise.[143] Søren Holm argues that the consultation was biased towards permitting non-medical sex selection for the purposes of family balancing.[144] He suggests that by using the term 'family balancing', the HFEA tried 'to create a very specific social reality in the hope that the reader will respond positively towards the plight of these poor unbalanced families'.[145] Further, he asserts that this bias can also be evidenced by the HFEA's separation of the reasons for choosing non-medical sex selection into 'personal reasons' and 'social/cultural/economic reasons'.[146] 'Personal reasons' appears to refer solely to family balancing, whereas 'social/cultural/economic reasons' includes having sons 'to work

[137] It is interesting to note that the Act appears to use the terms 'sex' and 'gender' interchangeably. Karpin notes that the use of the word gender 'is not insignificant', as determinations as to the 'seriousness of a condition… may also be contingent on gendered understandings of social norms of acceptable and challenging behaviour'. See Karpin I (2015), The Regulation of PGD for Medical Sex Selection and the Gendering of Disability in the UK and Australia. In: Horsey K (ed), *Revisiting the Regulation of Human Fertilisation and Embryology* (Routledge) 185, 188.

[138] Human Fertilisation and Embryology Authority, *Code of Practice* (9th ed, revised 2023) box 10D.

[139] Human Fertilisation and Embryology Authority, 'Sex Selection' (Public Consultation Document, January 1993).

[140] Human Fertilisation and Embryology Authority, 'Sex Selection: Choice and Responsibility in Human Reproduction' (Public Consultation Document, 2002) 6 [8].

[141] Both the 1993 and 2002 consultations looked at both pre-conception sex selection through sperm sorting and post-conception sex selection through PGT; however, this section will only discuss the use of PGT for sex selection.

[142] Holm S (2004) Like a Frog in boiling water: The Public, the HFEA and Sex Selection. 12 *Health Care Analysis* 27, 27.

[143] Ibid 28.

[144] Ibid 31.

[145] Ibid 32.

[146] Human Fertilisation and Embryology Authority, 'Sex Selection: Choice and Responsibility in Human Reproduction' (Public Consultation Document, 2002) 7.

the land' and avoiding the payment of dowries.[147] Holm notes that this dichotomy in turn creates an 'us' and 'them' divide in the mind of readers, whereby British people, particularly those of European descent, can identify with 'personal reasons', but may be more sceptical about 'social/cultural/economic reasons' that are often associated with people from other countries and ethnic backgrounds.[148]

This 'us' and 'them' divide appears again later in the consultation document, when quotes from the focus groups are used to illustrate various arguments for and against sex selection. While most quotes are labelled either mother or father, followed by age bracket and location, two quotes include the respondent's religion.[149] These quotes are labelled 'Muslim father, 25–40, London'. Holm questions why we are specifically told that these quotes come from a Muslim and asks why the other quotes do not contain equally informative labels.[150] The first of these quotes states that it is wrong to select the sex of a child 'for any reason. It's saying that you know better than what Allah wanted for you'. The second provides that '[h]aving a boy is like magic gold but it feeds discrimination'.[151] Holm suggests that one plausible explanation for this special label 'is that it is rhetorically important to label this expressed strong preference for sons as deviant from mainstream society'.[152]

The results of the 2002 consultation were the same as those of the 1993 consultation: that sex selection should only be permitted for medical reasons. The HFEA concluded that 'the likely benefits of permitting sex selection for non-medical reasons in the UK are at best debateable and certainly not great enough to sustain a policy to which the great majority of the UK public are strongly opposed'.[153] In coming to this decision, the HFEA relied upon the results of a MORI poll they had commissioned, which found that 69% of the people surveyed did not support the right of a parent to choose the sex of their future child.[154] This was further supported by the results of the questionnaire (found at the back of the consultation document), which showed 82.9% of respondents did not support the use of PGT for non-medical reasons.[155] When asked specifically about whether PGT should be permitted for family balancing, the proportion opposing sex selection remained virtually the same, 82.69%.[156] However,

---

[147] Ibid.

[148] Holm S (2004) Like a Frog in boiling water: The Public, the HFEA and Sex Selection. 12 *Health Care Analysis* 27, 34.

[149] Human Fertilisation and Embryology Authority, 'Sex Selection: Choice and Responsibility in Human Reproduction' (Public Consultation Document, 2002) 24–25.

[150] Holm S (2004) Like a Frog in boiling water: The Public, the HFEA and Sex Selection. 12 *Health Care Analysis* 27, 36.

[151] Human Fertilisation and Embryology Authority, 'Sex Selection: Choice and Responsibility in Human Reproduction' (Public Consultation Document, 2002) 24–25.

[152] Holm S (2004) 'Like a Frog in boiling water: The Public, the HFEA and Sex Selection' 12 *Health Care Analysis* 27, 36.

[153] Human Fertilisation and Embryology Authority, 'Sex Selection: Options for Regulation' (Report, 2003) 36 [147].

[154] Ibid 14 [45].

[155] Ibid 24.

[156] Ibid 25.

a higher proportion of respondents, 85%, opposed the use of sex selection for non-medical reasons other than family balancing (referred to in the consultation as 'social/cultural/economic reasons').[157]

Not only has criticism been directed at the HFEA's consultation document, but the HFEA's review of its findings has also been the subject of much debate. The main criticism of the HFEA's report is that it runs contrary to the presumption in favour of liberty; the idea that 'each citizen be granted the right to live his life as he sees fit, provided that in doing so he does not violate the rights of others'.[158] Edgar Dahl argues that the HFEA's conclusion that sex selection should only be permitted for medical reasons, in no small part based on the results of their poll and questionnaire, contravenes the presumption in favour of liberty.[159] He argues that the HFEA did not provide sufficient evidence of harm to justify a prohibition on non-medical sex selection.[160] Further, he notes that the fact that a majority of people oppose this kind of sex selection does not justify a prohibition in the absence of evidence of harm to society or the future child.[161] Thus in Dahl's view, to satisfy the presumption in favour of liberty, the HFEA would need to allow non-medical sex selection, wait for harm (such as significantly skewed sex ratios) to result, then use this evidence to prohibit the practice. Similarly, John Harris likens the HFEA's prohibition of non-medical sex selection to allowing a majority to determine what religion everyone should follow.[162] As this is not permitted, on the grounds that one's religion is a matter of personal choice, Harris contends that neither should an individual's reproductive choice as to the sex of their offspring be constrained without adequate justification.[163] Baldwin, on the other hand, distinguishes this kind of reproductive choice as one that is 'not fundamental to liberal democracy'.[164] Therefore, with regard to the use of PGT for non-medical sex selection, he suggests that public policy should be guided by public consensus.[165]

In a similar vein to Dahl and Harris, Guido Pennings asserts that had the HFEA, in the early 90s, conducted a public consultation on access to ART by lesbian couples, it is likely that a similar majority of the public would have voiced opposition, citing

---

[157] Ibid.

[158] Dahl E (2004) The Presumption In Favour of Liberty: A Comment on the HFEA's Public Consultation on Sex Selection. 8 *Reproductive Biomedicine Online* 266, 266.

[159] Ibid.

[160] See also Tizzard J (2004) Sex Selection, Child Welfare and Risk: A Critique of the HFEA's Recommendations on Sex Selection. 12 *Health Care Analysis* 61, 66–67.

[161] Dahl E (2004) The Presumption In Favour of Liberty: A Comment on the HFEA's Public Consultation on Sex Selection. 8 *Reproductive Biomedicine Online* 266, 266.

[162] Harris J (2005) Sex Selection and Regulated Hatred. 31 *Journal of Medical Ethics* 291, 293.

[163] Ibid.

[164] Baldwin T (2005) Reproductive Liberty and Elitist Contempt: Reply to John Harris. 31 *Journal of Medical Ethics* 288, 290.

[165] Ibid.

harm to future children.[166] He points out that studies have now shown that such concerns would have been unfounded.[167] As a consequence, Pennings suggests that it would be better for the HFEA to allow non-medical sex selection, strictly regulate it and then study its effect on the children born, their families and society as a whole.[168]

Further criticism of the HFEA's prohibition on non-medical sex selection has come from the House of Commons Science and Technology Committee. In its 2005 report, the Committee recommended that the 'onus should be on those who oppose sex selection for social reasons to show harm from its use'.[169] Thus to silence its critics, it appears that the HFEA needs to provide empirical evidence of the harm to children born as a consequence of non-medical sex selection. It may be possible for the HFEA to gather such empirical evidence from the US where non-medical sex selection is permissible.[170]

3.7 Selection for Disability

This section considers the United Kingdom's prohibition on the use of PGT for the purposes of selecting an embryo that will become a child with a disability. The use of PGT for these purposes is a kind of positive selection and is the antithesis of the kind of negative selection discussed earlier, i.e., selection against disability.

The *Human Fertilisation and Embryology Act 1990* (UK) did not contain specific references to selection in favour of disability, namely because it is unlikely that this use of PGT was contemplated at the time the Act was drafted. As discussed earlier in this chapter, schedule 2 paragraph 1(1)(d) of the 1990 Act provides that a treatment licence may be granted for the purposes of establishing 'that embryos are in a suitable condition to be placed in a woman'. As this section does not explicitly prohibit the use of PGT to select an embryo that will become a child with a disability, it has been suggested that the HFEA had the discretion to license such uses should they so choose.[171] While it is unlikely that the HFEA would have exercised such a discretion,

---

166 Pennings G (2004) 'Sex Selection, Public Policy and the HFEA's Role in Political Decision Making – Response to Edgar Dahl's 'The presumption in favour of liberty'' 8 *Reproductive Biomedicine Online* 268, 269.

167 Ibid.

168 Ibid. This suggestion has been echoed by John Robertson and John Harris. See Robertson JA (2004) Gender Variety as a Valid Choice: A Comment on the HFEA – Response to Edgar Dahl's 'The presumption in favour of liberty'. 8 *Reproductive Biomedicine Online* 270, 270; Harris J (2005) No Sex Selection Please, We're British. 31 *Journal of Medical Ethics* 286, 288.

169 Science and Technology Committee, *Human Reproductive Technologies and the Law*, House of Commons Paper No 7-I, Session 2004–05 (2005) 64 [142].

170 Ethics Committee of the American Society for Reproductive Medicine, 'Use of Reproductive Technology for Sex Selection for Nonmedical Reasons' (2015) 103 *Fertility and Sterility* 1418.

171 Porter G, Smith MK (2013) Preventing the Selection of "deaf embryos" under the Human Fertilisation and Embryology Act 2008: Problematizing Disability? 32(2) *New Genetics and Society* 171, 173.

the option for it to do so was removed by the *Human Fertilisation and Embryology Act 2008* (UK) s 14(4). This section inserts into s 13 of the 1990 Act a provision that prevents clinics from selecting embryos that are known to have a serious physical or mental disability, illness or medical condition, when embryos that are not affected are available for implantation. The Act does not prevent the implantation of an affected embryo in situations where all embryos carry the genetic disorder.

3.8 A Comparative Analysis of the Regulatory Frameworks for Embryo Testing in Australia and the UK

This section compares the regulatory frameworks for PGT in the UK and Australia. In developing a revised framework together with a set of underlying principles that may be better placed to respond to developments in embryo testing, this section asks whether there is anything Australia can learn from the UK experience.

The regulatory framework for PGT in the UK has been described as 'simpler and more coherent'[172] than the approach currently in place in Australia. Similarly, Brazier asserts that the British model has 'undoubted strengths',[173] including public accountability and the promotion of 'high standards of medical practice'.[174] While this level of coherence is easier to achieve in the UK due to the constitutional framework, a unified approach to the regulation of PGT across Australia would be desirable, particularly as the number of conditions and traits for which testing may be possible increases.[175] As the regulation of ART is a matter left to state and territory parliaments to legislate on, pursuant to s 51 of the Australian Constitution, the development of a federal legislative framework would require a referral from state and territory governments.[176] Such a referral would need the support of all eight states and territories for the resulting legislation to have national application.[177] In the alternative, legislation could be enacted using the corporations power in s 51(xx) of the Constitution.[178] This proposed legislation should give legislative support to the National Health and Medical Research Council Guidelines, which could then be used to elaborate upon

[172] Smith M (2007) Revisiting Old Ground in Light of New Dilemmas: The Need for Queensland to Reconsider the Regulation of Assisted Reproductive Technologies. 7 *Queensland University of Technology Law and Justice Journal* 425, 442.

[173] Brazier M (1998) Regulating the Reproduction Business? 7 *Medical Law Review* 166, 166.

[174] Ibid 177.

[175] This recommendation is further developed in Chap. 8.

[176] *Commonwealth of Australia Constitution Act 1901* (Cth) s 51(xxxvii).

[177] *Commonwealth of Australia Constitution Act 1901* (Cth) s 51(xxxvii) provides that the legislation created by a referral only applies to the states which referred the matter, and any states which later decide to adopt the law.

[178] *Commonwealth of Australia Constitution Act 1901* (Cth) s 51(xx) provides that the Commonwealth Parliament may pass legislation governing 'foreign corporations, and trading or financial corporations formed within the limits of the Commonwealth'.

and explain the provisions in the legislation, in much the same way as the HFEA Code of Practice supports the *Human Fertilisation and Embryology Act 1990* (UK).

In a similar vein, Commonwealth legislation could also establish a national regulatory body, similar to the HFEA. Therefore, rather than having multiple regulatory bodies, including the Fertility Society of Australia and New Zealand and the state regulators, a new body could be created that would take on all ART regulatory matters, in much the same way as the HFEA does. Alternatively, the Fertility Society of Australia and New Zealand could take on an expanded role. At present, even where regulatory bodies exist, the guidance they provide does not appear to be as comprehensive as that provided by the HFEA. For example, even prior to its dissolution the Victorian Assisted Reproductive Treatment Authority (VARTA) did not play an active role in the regulation of PGT, but oversaw compliance by reviewing the annual reports submitted by clinics. The benefit of a single regulatory body is that it will, in theory, result in policy-making that applies in a consistent manner across Australia. This in turn would make the regulatory framework in Australia more cohesive.

Some common themes emerge from both countries. Both Australia and the UK have facilitative frameworks that enable PGT to be carried out. In the UK and all Australian states, the use of PGT is permitted for the purposes of selection against disability. However, in the UK, Western Australia and South Australia, selection is limited to serious disabilities. Whether restricting the use of PGT to serious disabilities is desirable, or even sensible, is examined in Chap. 5. The UK and all Australian jurisdictions prohibit the use of PGT for non-medical sex selection. Similarly, all Australian jurisdictions and the UK prohibit using PGT to select an embryo that will become a child with a disability, when other embryos are available. The sanctioning of selection against disability while prohibiting selection for disability also suggests that both the Australian and UK regulatory frameworks may be premised on utilitarian principles of minimising harm, where the underlying assumption appears to be that being born with a disability is likely to negatively impact upon welfare.

Unlike the UK, in most Australian jurisdictions, with the exception of Western Australia where all new applications of PGT-M have to be approved by the Reproductive Technology Council, clinics may test for conditions for which PGT-M has not previously been utilised, without prior approval. Thus, one of the benefits of the Australian regulatory framework, is that in most states the decision as to when PGT-M may be used is made by clinics in consultation with prospective parents, within pre-set parameters, such as those contained in the National Health and Medical Research Council Guidelines. Conversely, in the UK, conditions for which PGT has not previously been used must be deemed 'sufficiently serious' by the HFEA. This additional level of scrutiny seems unnecessary, particularly when clear guidance for clinics exists, with consequences for deliberate breaches.

With regard to HLA typing, all Australian jurisdictions and the UK allow the use of HLA typing in conjunction with PGT. Further, with the exception of Western Australia, all jurisdictions allow HLA typing to be used in situations where PGT is not required. This is a further example of the lack of consistency that exists across Australia. A point of difference between the two countries is that Australia permits the use of HLA typing for the creation of a 'saviour relative', whereas HLA typing in the UK is restricted to siblings.

Australia and the UK both have facilitative frameworks for the regulation of embryo testing. These frameworks are underpinned by utilitarian principles of welfare maximization. As a consequence, the regulatory regimes operating in both countries permit embryo testing for the purposes of selection against disability and HLA typing. The *Human Fertilisation and Embryology Act 1990* (UK) established the HFEA and this section has highlighted why Australia would benefit from a national regulator, not just for embryo testing, but for ART as a whole.

3.9 Conclusion

This chapter has aimed to provide an overview of the regulatory framework for PGT in the UK. It has discussed the origins of ART regulation, as well as the 2008 reforms. The *Human Fertilisation and Embryology Act 2008* (UK) provided a much-needed update to the *Human Fertilisation and Embryology Act 1990* (UK). However, the 2008 Act has been criticised for doing little more than codifying what was already happening.[179] Nevertheless, it is important that legislation reflects social norms, for example in areas such as parenthood and eligibility for treatment, and the 2008 reforms reflect the fact that the heteronormative family form is no longer the norm for many families in the UK.

The *Human Fertilisation and Embryology Act 1990* (UK) permits the use of PGT for the purpose of selection against disabilities and also allows sex selection for the avoidance of sex linked disorders. However, PGT is limited to conditions that are classed as serious by the HFEA's SAC. This creates two problems. First, it creates an unnecessary level of bureaucracy, whereby decisions are made by the SAC as opposed to clinics in consultation with prospective parents. Second, determinations as to seriousness are often a matter of perspective and it is questionable whether objective determinations of seriousness in the context of disability are helpful or even possible.

The only kind of positive selection permitted by the *Human Fertilisation and Embryology Act 1990* (UK) is HLA typing for the selection of a saviour sibling. Other kinds of positive selection, such as non-medical sex selection and selection for disability, are not allowed under the 1990 Act. The restrictions placed on positive selection in the UK are the same as those in Australia. While these kinds of positive selection should be prohibited, in the following chapters I argue that regulatory frameworks should be open to permitting other kinds of positive selection once the technology becomes available. In particular, selection that furthers the interests of the future child, as opposed to those of the parents.

[179] Snelling J, Gavaghan C (2015) PGD Past and Present: Is the HFE Act 1990 Now 'Fit for Purpose'? In: Horsey K (ed) Revisiting the Regulation of Human Fertilisation and Embryology (Routledge) 80, 93.

In reconsidering the regulatory framework for embryo testing in Australia, there are a number of lessons that can be learned from the UK regulatory experience. Australia may benefit from a national regulatory framework to ensure consistency across all states and territories.[180] This framework could take the form of Commonwealth legislation. If legislation is enacted, it should provide legislative backing to the NHMRC Guidelines. The Guidelines could then serve the same purpose as the HFEA's Code of Practice.

In the chapters that follow, each aspect of the regulation of embryo testing will be examined in order to arrive at a decision on what the parameters of a revised framework should be.

---

[180] This is discussed in Chap. 8.

Chapter 4
Theoretical Underpinnings: Disability Theory and Theories of Reproductive Liberty

Abstract This chapter discusses the theory behind two concepts that play a central role in the chapters which follow. The first of these concepts is that of disability. This chapter considers some of the ways in which disability has been modelled, with the aim of selecting a model capable of application in the sphere of reproductive choice. The second concept discussed is that of reproductive liberty. In a similar vein to theories of disability, a selection of approaches to reproductive liberty will be discussed. One theory will be selected as the basis for discussions around the limits of reproductive choice in the context of embryo testing.

4.1 Introduction

The preceding chapters provided an overview of the regulatory frameworks for embryo testing in Australia and the United Kingdom. They did so in order to ascertain the elements of those frameworks that should be retained and those that can be improved in the development of a revised framework for embryo testing. The aim of this chapter is to set out the theoretical underpinnings of two aspects of the forthcoming chapters. This chapter first sets out various models of disability. It does so with a view to selecting a model of disability that will form the basis of a framework for decisions around reproductive choice in the context of both selection against disability and selection for disability. This section will begin by considering the two traditional models of disability: the medical model and the social model. The discussion will then turn to more recently developed models of disability such as the interactional model, the welfarist model, and the World Health Organization's biopsychosocial model, among others.

Following on from this, the chapter looks at theories of reproductive liberty. It asks, what is the right protected by this concept and why is it important? A number of academics have developed their own theories of reproductive liberty and a selection of these will be discussed. Outlining the theoretical basis of reproductive liberty assists in answering the question which is central to the forthcoming chapters, namely, what kind of children should prospective parents have the freedom to create?

M. de Souza, *The Regulation of Embryo Testing in Australia*,
https://doi.org/10.1007/978-981-96-5902-9_4

4.2 What is Disability?

Defining and theorizing disability is complex due to the diversity of the conditions that fall within the broad concept of disability.[1] As Margrit Shildrick observes, what is considered a disability 'varies greatly according to the socio-historical and geopolitical context, and even in a single location the designation remains stubbornly multifaceted and resistant to definition in terms of both its boundaries and meanings'.[2] In spite of this difficulty, a number of disability scholars, disability activists and bioethicists have developed models of disability.[3] There are two dominant models of disability: the medical model and the social model. This section begins by discussing these models, before considering more contemporary models of disability.

4.2.1 The Medical Model of Disability

The medical model of disability conceptualises disabilities as harmful conditions that emanate from within a person. As Jackie Leach Scully notes, 'from a medicalized perspective, disability is an abnormality of form or function, the cause of which lies in the biology of the individual'.[4] Mike Oliver, one of the first theorists to discuss the modelling of disability,[5] notes that there are two important features of the medical model. The first is that disability is located within the individual. The second is that it attributes the problems caused by disabilities to physical or psychological limitations.[6] Thus for Oliver, a significant part of the medical model, or as he refers

---

[1] Asch A, Wasserman D (2014) Reproductive Testing for Disability. In: Arras J D, Fenton E, Kukla R (eds) The Routledge Companion to Bioethics. Routledge, New York, p 417–432, 418; Scotch R K, Schriner K (1997) Disability as Human Variation: Implications for Policy. The Annals of the American Academy of Political and Social Science, 549(1), 148–159, 154; Scully JL (2008) Disability and genetics in the era of genomic medicine. Nature Reviews Genetics, 9(10), 797–802, 800.

[2] Shildrick M (2012) Critical Disability Studies: Rethinking the Conventions for the Age of Postmodernity. In: Watson N et al. (eds) The Routledge Handbook of Disability Studies. Taylor & Francis, Florence, p 30–41, 33.

[3] The issue is further complicated by related concepts such as disease. Disability and disease are separate concepts, although they are often conflated (Scully JL (2004) What is a Disease? EMBO Reports 5(7) 650–653, 653). Defining disease is complicated for similar reasons to defining disability, however, as Scully notes 'If defining disease is difficult, disability is worse'. The use of the term disability in this book encompasses disease. For a detailed discussion of disease, see Kukla QR (2022) What Counts as a Disease and why does it matter? Journal of Philosophy of Disability 2: 130–156.

[4] Scully JL (2008) Disability Bioethics: Moral Bodies, Moral Difference. Rowman & Littlefield, 28.

[5] Oliver M, *Understanding Disability: From Theory to Practice* (Macmillan Education UK, 1996) 30.

[6] Oliver M, *Understanding Disability: From Theory to Practice* (Macmillan Education UK, 1996) 32.

to it, the individual model, is the 'medicalisation' of disability.[7] Consequently, the medical model focuses on preventing and treating disability.[8]

As may be expected, even within the medical model different definitions of disability can be found, as illustrated by those put forward by Jonathan Glover and John Harris. Glover defines disability as a 'failure or limitation of functioning' which on its own or in conjunction with social disadvantage 'impairs the capacity for human flourishing'.[9] Therefore, where a person's condition is not limiting and no medical complications are associated with it, then that condition should not be classed as a disability.[10] Glover illustrates this with the example of a person with a fictional medical condition that results in their toenails not growing.[11] He contends that as the only symptom of this condition is a lack of toenail growth, it should not be considered a disability. This is because no functional limitation is associated with toenails not growing and thus it does not impair the person's capacity for flourishing.[12] For Glover, whether or not a limitation of functioning is classed as a disability is assessed by reference to a standard of normality, or 'normal human functioning'.[13] This contrast with normality has been criticised by John Harris who points out that '[n]ormal species functioning cannot form part of the definition of disability because people might be normal and still disabled'.[14] Thus Harris defines disability as a 'harmed condition' and one that 'someone has a strong rational preference not to be in'.[15] It is evident from this definition that Harris considers that what counts as a disability should be considered 'relative to possible alternatives'.[16] This is, however, problematic as it casts 'the net too wide'[17] and could, by extension, mean that anyone with a strong rational preference to be in an alternative state, for example, more intelligent, would be classified as disabled.[18] While it is true that higher levels of intelligence may increase capacities for flourishing, it seems absurd to say that anyone who has a rational desire to be more intelligent could, as a result, be classed as disabled. To use Glover's expression, Harris' 'rational reconstruction is at the cost of great detachment from the ordinary concept of disability'.[19]

---

[7] Oliver M, *Understanding Disability: From Theory to Practice* (Macmillan Education UK, 1996) 31.

[8] Oliver M, *Understanding Disability: From Theory to Practice* (Macmillan Education UK, 1996) 36. See generally Haegele JA, Hodge S (2016) Disability Discourse: Overview and Critiques of the Medical and Social Models. Quest 68(2): 193–206, 195.

[9] Glover J, *Choosing Children: Genes, Disability and Design* (Oxford University Press, 2006) 9.

[10] Glover J, *Choosing Children: Genes, Disability and Design* (Oxford University Press, 2006) 10.

[11] Ibid 9.

[12] Ibid.

[13] Ibid 11.

[14] Harris J, 'One Principle and Three Fallacies of Disability Studies' (2001) 27 *Journal of Medical Ethics* 383, 384.

[15] Ibid.

[16] Ibid.

[17] Glover J, *Choosing Children: Genes, Disability and Design* (Oxford University Press, 2006), 13.

[18] Ibid.

[19] Ibid.

The medical model of disability has been criticised for necessarily associating disability with disadvantage.[20] Its critics argue that even though a person has a disability, that is not to say that they are in any way harmed and in need of medical treatment to 'fix' them.[21] For example, Fiona Kumari Campbell criticises the medical model for feeding 'into unquestioned ableist assumptions'.[22] She defines ableism as a system of 'beliefs, processes and practices' that promote a species-typical image of the body as 'fully human' while casting disability 'as a diminished state of being human'.[23] Thus a key element of ableism is 'a belief that impairment or some adversely differentiated attribute is inherently negative and at its essence is a form of harm in need of amelioration, cure or indeed exculpation.'[24] Therefore, for Campbell, the medical model's focus on curing disability reflects ableist beliefs about the nature of disability.

One of the reasons for the medical model's dominance is its prevalence in the fields of law and medicine. In the legal sphere, it forms the basis of legislation such as the *Disability Discrimination Act 1992* (Cth). This Act defines disability in a number of ways including

(a) total or partial loss of the person's bodily or mental functions; or
(b) total or partial loss of a part of the body; or
(c) the presence in the body of organisms causing disease or illness; or
(d) the presence in the body of organisms capable of causing disease or illness; or
(e) the malfunction, malformation or disfigurement of a part of the person's body.[25]

The medical model's ideology is evident in the definition of disability as a person's physical or mental loss, or bodily malfunction.[26]

As the medical model represents the dominant view in medicine, it is used as the basis for resource allocation within health care, giving rise to measures such as quality adjusted life years (QALYs) and disability adjusted life years (DALYs). According to the QALYs approach, a year during which a person is healthy is given

---

[20] Koch T, 'Disability and Difference: Balancing Social and physical constructions' (2001) 27 *Journal of Medical Ethics* 370.

[21] Marks D (1999), *Disability: Controversial Debates and Psychosocial Perspectives* (Routledge) 4.

[22] Campbell FK (2009) Contours of Ableism: The Production of Disability and Abledness. Palgrave MacMillan, 98.

[23] Campbell FAK (2001) Inciting Legal Fictions: Disability's Date with Ontology and the Ableist Body of the Law. Griffith Law Review 10(1): 42–62, 44.

[24] Campbell FK (2019) Precision Ableism: A *Studies in Ableism* Approach to Developing Histories of Disability and Abledment. Rethinking History 23(2): 138–156, 148.

[25] *Disability Discrimination Act 1992* (Cth) s 4 (definition of 'disability').

[26] The dominance of the medical model in disability discrimination statutes can also be seen in other jurisdictions, e.g. the *Americans With Disabilities Act of 1990*. Donoghue notes that in the American context, while disability advocates welcomed the advent of the *Americans With Disabilities Act* as a way of improving their lives, the model of disability conceptualised in the Act was very different to that for which they were advocating. See Donoghue C (2003) Challenging the Authority of the Medical Definition of Disability: An Analysis of the Resistance to the Social Constructionist Paradigm. *Disability & Society* 18(2) 199–208, 203.

a value of one, whereas a year of ill health is valued at less than one, on a sliding scale where death equals zero. Using this method, a calculation can be undertaken to ascertain whether the cost of a specific treatment is proportional to the benefits of that treatment. For example, if a person has a severe disability valued at 0.2 and requires an expensive treatment that will not cure their disability but will increase their life expectancy from five to ten years, the calculation of the QALYs would be $0.2 \times 10 = 2$ QALYs; without treatment, the person would have 1 QALY. It would therefore be difficult to justify an expensive treatment that yielded such a small number of QALYs.[27]

This overview of the medical model of disability has highlighted that although proponents of the medical model may differ in their definitions of disability, there is general agreement as to the features at its core. The key elements of this model are first, that disabilities emanate from within a person and second, that they are negative in nature.

4.2.2 The Social Model of Disability

The idea behind the social model of disability stemmed from the work of the Union of the Physically Impaired Against Segregation (UPIAS) in the UK.[28] In 1976, the UPIAS described disability as 'the disadvantage or restriction of activity caused by a contemporary social organisation which takes little or no account of people who have physical impairments and thus excludes them from participation in the mainstream

[27] DALYs work in a similar way to QALYs, by assigning a disability weighting to states of ill health, with perfect health having a weighting of zero and death a weighting of one. It has been suggested that the 'DALY approach explicitly presupposes that the lives of disabled people have less value than those of people without disabilities' and 'assumes that disabled people are less entitled to scarce health resources for interventions that would extend their lives' (Arnesen T, Nord E (1999), The Value of DALY Life: Problems with Ethics and Validity of Disability Adjusted Life Years. 315 *British Medical Journal* 1423, 1423). These criticisms apply equally to QALYs, which have also been criticised for discriminating against elderly people, who are unlikely to be able to compete with younger patients in terms of the number of years they have left (Newdick C (2005), *Who Should We Treat? Rights, Rationing and Resources in the NHS* (Oxford University Press) 29). For a detailed critique of DALYs, see Anand S, Hanson K, 'Disability-Adjusted Life Years: A Critical Review' in Anand S, Peter F, Sen A (eds), *Public Health, Ethics, and Equity* (Oxford University Press, 2004) 183.

[28] Oliver M (2013) The Social Model of Disability: Thirty Years On. *Disability & Society* 28(7): 1024–1026, 1024; Oliver M, *Understanding Disability: From Theory to Practice* (Macmillan Education UK, 1996) 30–31. Richard Light notes that it is possible to trace the origins of the social model back to the 1960s and the work of Paul Hunt, a Briton with muscular dystrophy. Light notes that Hunt's essay 'A Critical Condition' was the first work to describe disability in terms of the social barriers inflicted upon people with impairments: Light R (2010) Social Model or Unsociable Muddle? In: Rix J, Nind M, Sheehy K, Simmons K, Walsh C (eds) Equality, Participation and Inclusion 1: Diverse Perspectives. Routledge, Florence, 119–123.

of social activities'.[29] Richard Light notes that even though the UPIAS description specifically refers to physical impairments, the insights offered by the social model were soon considered to be valuable to all people with disabilities.[30] Proponents of the social model of disability argue that disability is located outside the body and is thus a social issue. Therefore, the issues experienced by people with disabilities are due to social barriers and discriminatory attitudes. Locating the nature of disability in society places the social model in stark contrast to the medical model, which has been said to 'individualise the nature of disability'.[31] Thus, the social model

> signals that the experience of disabled people is dependent on social context, and differs in different cultures and at different times. Rather than disability being inescapable, it becomes a product of social arrangements and can thus be reduced, or possibly even eliminated.[32]

A notable feature of this model is the distinction it draws between disability and impairment. Disability is a social construct, whereas impairment refers to biological factors, such as cognitive impairments or physical impairments.[33] Therefore, according to the social model, impairment and disability are not synonymous. A person may have an impairment, such as a physical impairment that requires the use of a wheelchair, but in a wheelchair accessible society this impairment is not a disability. As Iris Marion Young asserts, 'Moving on wheels is a "disadvantage" only in a world full of stairs'.[34] This highlights the role of culture in defining disability within the social model. Owing to this dichotomy, whether or not a person has a disability will depend on the extent to which social barriers and discriminatory attitudes prevail in their culture.[35] As a result, a person living in a culture in which accessibility is a high priority may not be disabled, whereas the same might not be said for someone with an identical impairment living in a society with numerous social barriers. It should, however, be noted that the impairment–disability distinction is not a universal tenet of the social model. As Anita Silvers observes, 'There is nothing

---

[29] UPIAS, quoted in Light R (2010) Social Model or Unsociable Muddle? In: Rix J, Nind M, Sheehy K, Simmons K, Walsh C (eds) Equality, Participation and Inclusion 1: Diverse Perspectives. Routledge, Florence, 119–123, 120.

[30] Light R (2010) Social Model or Unsociable Muddle? In: Rix J, Nind M, Sheehy K, Simmons K, Walsh C (eds) Equality, Participation and Inclusion 1: Diverse Perspectives. Routledge, Florence, 119–123, 121.

[31] Donoghue C (2003) Challenging the Authority of the Medical Definition of Disability: An Analysis of the Resistance to the Social Constructionist Paradigm. *Disability & Society* 18(2) 199–208, 207.

[32] Shakespeare T, *Disability Rights and Wrongs* (Routledge, 2006) 29.

[33] As Oliver notes, impairment is simply 'a description of the physical body'—Oliver M, *Understanding Disability: From Theory to Practice* (Macmillan Education UK, 1996) 35.

[34] Young IM, 'Foreword' in Mairian Corker and Tom Shakespeare (eds), *Disability/Postmodernity: Embodying Disability Theory* (Continuum, 2002) xii.

[35] Goodley D, Hughes B, Davis L (2012) Introducing Disability and Social Theory. In: Goodley D, Hughes B, Davis L (eds), *Disability and Social Theory* (Palgrave Macmillan) 1. This point has also been made by Jasbir Puar, who suggests that the divide between disability and ability 'traverses social, geographic and political spaces': Jasbir Puar, *The Right to Maim: Debility, Capacity, Disability* (Duke University Press, 2017) xiv.

about social model theory… that entails or otherwise calls for this dichotomization. Indeed, impairment itself is not a fixed standard'.[36] Thus she argues that impairment is also socially constructed (as opposed to biological fact) as it is based on the premise that a particular physical or cognitive state differs from what is considered 'normal'.[37]

Even within the social model different theoretical perspectives exist, such as the minority group model and the human variation model.[38] The minority group model views people with disabilities as a minority group (similar to an ethnic minority) and the disadvantages they face are due to stigma and discrimination.[39] On the other hand, the human variation model regards disability as an aspect of human diversity and therefore as something that may affect everyone at some point during their lifetime; as a result, it has been referred to as a 'universalist' approach.[40] The human variation model considers the difficulties faced by people with disabilities to stem from a lack of congruence between their impairments and social structures.[41] Thus it sees society's lack of willingness to accommodate human variation as the source of the problems associated with disability.[42] Supporters of the human variation model argue that it is preferable to the minority group model, as moving away from the problems associated with discrimination allows for the development of a broader range of solutions in areas including medicine, education and human services.[43] However, as Asch and Wasserman observe, even though the minority group model and the human variation model differ in their focus, their goals overlap, as the consequences of discrimination can be seen in society's failure to accommodate human difference.[44]

[36] Silvers A (2010), 'An Essay on Modeling: The Social Model of Disability' in D Christopher Ralston and Justin Ho (eds), *Philosophical Reflections on Disability* (Springer) 19, 24.

[37] Ibid. A similar point is made by Lennard Davis who notes that 'the impairment diagnosis itself is not exempt from the forces of social construction and ideological control'—Davis L (2013) The End of Normal: Identity in a Biocultural Era. University of Michigan Press, Ann Arbor, 62.

[38] Asch A, Wasserman D (2014). Reproductive Testing for Disability. In: Arras J D, Fenton E, Kukla R (eds) The Routledge Companion to Bioethics. Routledge, New York, p 417–432, 419.

[39] Ibid.

[40] Gooding P, Kayess R (2022) Human Rights and Disability: An Australian Experience. In Gerber P, Castan M (eds) Critical Perspectives on Human Rights Law in Australia. Thomson Reuters, Sydney, 191–214, 193.

[41] Asch A, Wasserman D (2014). Reproductive Testing for Disability. In: Arras J D, Fenton E, Kukla R (eds) The Routledge Companion to Bioethics. Routledge, New York, p 417–432, 419.

[42] Scotch R K, Schriner K (1997) Disability as Human Variation: Implications for Policy. The Annals of the American Academy of Political and Social Science, 549(1), 148–159, 155.

[43] Ibid 156.

[44] Asch A, Wasserman D (2014). Reproductive Testing for Disability. In: Arras J D, Fenton E, Kukla R (eds) The Routledge Companion to Bioethics. Routledge, New York, p 417–432, 419. For a detailed discussion of the differences between the minority group model and the human variation model (which she refers to as the universal model) see Goldiner A (2022) Understanding Disability as a Cluster of Disability Models. Journal of Philosophy of Disability 2: 28–54, 46–48. https://doi.org/10.5840/jpd20224411.

Some social model theorists have suggested that disability is a neutral concept, as opposed to a negative one, in much the same way as race is a neutral concept.[45] Accordingly, having a disability can be disadvantageous in a non-inclusive society in much the same way that belonging to a minority race is a disadvantage in a racist society. Therefore, as race is not considered a harm, neither should disability. However, this ideal has been met with criticism. Anastasiou and Kauffman argue that

> disabilities themselves are not neutral features... Despite the fact that disability is part of human diversity, it is not just another difference and cannot be equated with social disadvantage... there are important differences among differences and disability is not a mere social construct.[46]

Further, they note that relatively few of the difficulties faced by people with disabilities 'can be attributed *exclusively* to social barriers'.[47] This can be illustrated by John Harris' HIV/AIDS analogy.[48] Writing in 2000, Harris argued that although reducing the stigma of HIV/AIDS may improve the lives of those living with the condition, it would not address the compromised immune system of people with HIV/AIDS.[49] Therefore, while social factors can make the lives of people living with HIV/AIDS less difficult, they do not assist with the biological elements of the conditions.

Advocates of the social model have expressed resistance to the concept of curing or attempting to cure disability, because by attempting to cure impairments, which are not themselves limiting, individuals are being 'corrected to fit in with the norm'.[50] However, as Tom Shakespeare points out, it would be 'misguided to oppose cures and therapies in principle',[51] because even in cases where individuals can function adequately, medical research could prevent further loss of function. Supporters of the social model, on the other hand, have argued that funds should be directed away from research that aims to cure impairments and towards the removal of barriers to social inclusion.[52] Conversely, Shakespeare argues that '[t]he goal of promoting cultural respect and social acceptance for people with impairment should not distract us from

---

[45] See generally, Reindal SM (2000) ,Disability, Gene Therapy and Eugenics—a challenge to John Harris. 26 *Journal of Medical Ethics* 89; Silvers A, 'On the possibility and desirability of constructing a neutral conception of disability' (2003) 24 *Theoretical Medicine* 471; Koch T, 'Disability and Difference: Balancing Social and Physical Constructions' (2001) 27 *Journal of Medical Ethics* 370.

[46] Anastasiou D, Kauffman J, 'The Social Model of Disability: Dichotomy between Impairment and Disability' (2013) 38 *Journal of Medicine and Philosophy* 441, 446.

[47] Anastasiou D, Kauffman J, 'Disability as Cultural Difference: Implications for Special Education' (2012) 33(3) *Remedial and Special Education* 139, 143.

[48] Harris J, 'Is there a Coherent Social Conception of Disability?' (2000) 26 *Journal of Medical Ethics* 95, 99.

[49] Ibid.

[50] Shakespeare T, *Disability Rights and Wrongs* (Routledge, 2006) 108.

[51] Ibid 109.

[52] See, eg, Oliver M, *Understanding Disability: From Theory to Practice* (Macmillan Education UK, 1996) 100–101.

the importance of mitigating or preventing impairment'.[53] As Susan Wendell writes, even those who strongly identify 'with being disabled and have very happy, fulfilling lives might want to be cured, not for social reasons, but because they want to have less pain or physical or mental difficulty'.[54]

Critics of the social model argue that even if all social barriers and discriminatory attitudes were removed, individuals would still have impairments that would prevent them from flourishing. For example, Harris points out that it is not

> social factors that make blindness and lameness and deafness into a disability. Social factors may exacerbate the problem of having such disabilities but they are disabilities because there are important options and experiences that are foreclosed by lameness, blindness and deafness.[55]

Similarly, Shakespeare notes that the 'problems associated with disability cannot be eliminated by any imaginable form of social arrangements. The priority for a progressive disability politics is to engage with impairment, not to ignore it'.[56] This lack of engagement with impairment gives rise to one of the key weaknesses of the social model; that it does not sufficiently take into account the subjective nature of disability. For example, for someone with an impairment that causes them significant pain, the removal of social barriers does not reduce the pain associated with their condition. Therefore, by failing to adequately consider 'the actual or potential personal impact of impairment(s)' the social model neglects a fundamental element of disability.[57] As Shakespeare points out, the 'social model so strongly disowns individual and medical approaches, that it risks implying that impairment is not a problem'.[58] Further, he argues that particularly for those with conditions that cause pain and suffering, 'it is harder to ignore the negative aspects of impairment'.[59] In a similar vein, Jenny Morris notes that in their eagerness to move away from the medical model of disability, people with disabilities 'have sometimes tended to deny the personal experience of disability'.[60] Interestingly, Oliver, one of the original proponents of the social model, asserts that when the social model was first developed, it was neither meant to replace the medical model, nor act 'as an all-encompassing framework within which everything that happens to disabled people could be understood or explained'.[61] Rather, the aim of the social model was to

---

[53] Shakespeare T, *Disability Rights and Wrongs* (Routledge, 2006) 116.

[54] Wendell S (1996) The Rejected Body: Feminist Philosophical Reflections on Disability. Routledge, London, 82–83.

[55] HarrisJ, 'Is there a Coherent Social Conception of Disability?' (2000) 26 *Journal of Medical Ethics* 95, 98.

[56] Shakespeare T, *Disability Rights and Wrongs* (Routledge, 2006) 56.

[57] McLean S, Williamson L, *Impairment and Disability: Law and Ethics at the Beginning and End of Life* (Routledge, 2007) 24.

[58] Shakespeare T., 'The Social Model of Disability' in Lennard Davis (ed), *The Disability Studies Reader* (Routledge, 2nd ed, 2006) 197, 200.

[59] Ibid.

[60] Morris J (1991) *Pride Against Prejudice* (The Women's Press) 181.

[61] Oliver M (2013) The Social Model of Disability: Thirty Years On. *Disability & Society* 28(7): 1024–1026, 1024.

encourage professionals working with people with disabilities to 're-orient' their work to make it 'more relevant to the needs of disabled people'.[62]

The social model occupies an important place in disability studies, as it is 'symbolic' of the relationship between disability studies as an academic discipline 'and its social roots—the disabled people's movement'.[63] However, as Carol Thomas notes, what is needed is greater 'constructive theoretical engagement with the significance of impairment and impairment effects in the lives of disabled people'.[64] Thus Thomas suggests that a 'social relational theory' of disability is needed to build upon the work that has been done by the social model.[65]

4.2.3 The Interactional Model

As the above discussion indicates, both of the traditional models of disability present difficulties and this has prompted some academics and organisations to develop their own approaches to defining disability, incorporating what they see as the beneficial elements of each model. The first model to be examined is the Interactional Model, put forward by Tom Shakespeare. The interactional model considers disability 'as an interaction between individual and structural factors'.[66] Therefore, rather than adopting the medical model's ideology that disability emanates from within a person, or the social model's perspective that disability is created by society, Shakespeare argues that 'people are disabled by society and by their bodies'.[67] According to Shakespeare, the interactional model is preferable to both the social and medical models as it is capable of encompassing the diverse range of impairments experienced by people with disabilities.[68] Indeed an important feature of the interactional model is the way in which it views impairment. Shakespeare considers impairment as a predicament because calling 'something a predicament is to understand it as a difficulty, and as a challenge, and as something which we might want to minimise but which we cannot ultimately avoid'.[69]

Shakespeare acknowledges that some may dispute his categorisation of impairment as a negative construct, however, he points out that if impairment was in

[62] Ibid.

[63] Thomas C (2004) Rescuing a Social Relational Understanding of Disability. Scandinavian Journal of Disability Research 6(1): 22–36, 33–34. https://doi.org/10.1080/15017410409512637.

[64] Thomas C (2004) Rescuing a Social Relational Understanding of Disability. Scandinavian Journal of Disability Research 6(1): 22–36, 33. https://doi.org/10.1080/15017410409512637.

[65] Ibid.

[66] Shakespeare T, *Disability Rights and Wrongs* (Routledge, 2006) 55.

[67] Ibid 56.

[68] Ibid 60.

[69] Ibid 63.

reality neutral, then to deliberately impair a child would be acceptable.[70] As it is highly unlikely that anyone would consider this practice moral, he argues that this implies that preventing impairment 'must have an important role in social responses to disability'.[71] As Shakespeare notes, 'impairment almost always plays some role in the lives of disabled people, even if social arrangements or cultural context minimises the exclusion or disadvantage'.[72] By considering disability in this way, the interactional model takes into account the realities of those with severe disabilities for whom removal of social barriers is of little use due to the severity of their impairment. Further, the interactional model leaves open the option to combine a range of quality of life enhancing options, thereby providing people with disabilities with a combination of therapies, physical aids and environmental changes to suit their individual needs.[73]

4.2.4 *The Welfarist Account*

An alternative model of disability, termed the Welfarist Account, has been proposed by Guy Kahane and Julian Savulescu. This model defines disability as a 'stable physical or psychological property of subject S that leads to a reduction of S's level of wellbeing in circumstances C'.[74] Therefore, a biological or psychological condition is a disability if it increases the likelihood 'that a person's life will get worse, in terms of his or her own wellbeing, in a given set of social and environmental circumstances'.[75]

One facet of this model is that it does not assess what constitutes a disability by reference to normality.[76] This is because whether or not 'an intrinsic property of a person'[77] reduces their wellbeing is not dependent upon whether that property deviates from the norm. As a result, personality traits such as poor impulse control or recklessness, could on the welfarist approach, count as disabilities.[78] They also suggest that achondroplasia (dwarfism) would, at most, fall into the category of

---

[70] Ibid 64. Elizabeth Barnes argues that this line of thinking is flawed, see Barnes E (2014) Valuing Disability, Causing Disability. Ethics 125(1): 88–113. This issue is explored further in Chap. 7.

[71] Shakespeare T, *Disability Rights and Wrongs* (Routledge, 2006) 64.

[72] Ibid 61.

[73] Ibid 62.

[74] Kahane G, Savulescu J (2009) 'The Welfarist Account of Disability' in Brownlee K, Cureton A (eds), *Disability and Disadvantage* (Oxford University Press) 14, 25.

[75] Savulescu J, Kahane G, 'Disability: A Welfarist Approach' (2011) 6 *Clinical Ethics* 45.

[76] Kahane and Savulescu, 'The Welfarist Account of Disability' in Kimberley Brownlee and Adam Cureton (eds), *Disability and Disadvantage* (Oxford University Press, 2009) 14, 26.

[77] Ibid.

[78] Ibid.

mild disability, as being short is less likely to reduce wellbeing than the personality traits mentioned above.[79] However, this fails to acknowledge the other symptoms of achondroplasia, such as spinal stenosis,[80] that may cause back pain, cramping and numbness, and sufferers may have difficulty walking for long periods.[81] Further, Kahane and Savulescu note that deafness can be both a disability and a neutral characteristic. It may be considered a disability as it reduces wellbeing 'by preventing access to the world of sound'.[82] Alternatively, deaf people who are immersed in large deaf communities are less likely to be classed as disabled because the world of sound does not play a significant role in their lives and thus does not reduce their wellbeing. As Kahane and Savulescu observe, it 'is the exercise of the capacity to hear that is valuable, not the capacity itself'.[83]

According to the welfarist account, disability is context-dependent. Consequently, to classify a condition as disabling, consideration needs to be given to the person who has or will have the condition and the environment in which they will exist.[84] As Michael Dunn points out, social situations in which a person has a lower quality of life due to their physical or psychological properties would count as disabling circumstances.[85] Therefore, to determine whether a person has a disability necessitates 'making judgements about that person's wellbeing through observing who that person is, and how she lives her life'.[86]

Elizabeth Barnes describes the welfarist account as 'radically revisionary'.[87] This is because, as Kahane and Savulescu themselves note, on the welfarist account certain personality traits would count as disabilities. Barnes, on the other hand, suggests that a successful model of disability should 'deliver correct verdicts for paradigm cases'.[88] Therefore, an account that considers traits such as recklessness as a disability fails to meet this criterion. For Barnes, a 'theory of disability shouldn't be in the business of claiming that much of what is said about disability within the disability rights movement is analytically false'.[89]

---

[79] Ibid 50.

[80] Beaudreuil J et al., 'Lumbar Spinal Stenosis in Adult Achondroplasia: An Analysis of Intervertebral Disk Alterations' (2018) 61 *Annals of Physical and Rehabilitation Medicine* e160.

[81] Waldman SD, 'Spinal Stenosis' in Steven D Waldman (ed), *Atlas of Common Pain Syndromes* (Elsevier, 4th ed, 2019) 324, 325.

[82] Kahane and Savulescu, 'The Welfarist Account of Disability' in Brownlee K, Cureton A (eds), *Disability and Disadvantage* (Oxford University Press, 2009) 14, 49.

[83] Ibid.

[84] Ibid 27.

[85] Dunn M, 'Discourses of Disability and Clinical Ethics Support' (2011) 6 *Clinical Ethics* 32, 35.

[86] Ibid.

[87] Barnes E (2016) The Minority Body: a theory of disability. Oxford University Press, Oxford 12.

[88] Ibid 10.

[89] Ibid 12.

Use of the welfarist account is problematic in the context of reproductive choices as it is difficult to make a context-dependent assessment of whether a condition is disabling in relation to a person who has not been born. It may, however, be possible to make a projected assessment of disability by analysing the future child's genetic condition in light of the most likely circumstances into which the child will be born.

4.2.5 *The Biopsychosocial Model*

The next model to be considered is the World Health Organisation's (WHO) biopsychosocial model, which forms the basis for the International Classification of Functioning, Disability and Health (ICF). In a similar vein to the interactional model, the biopsychosocial model views disability as 'an interaction between features of the person and features of the overall context in which the person lives'.[90] According to the WHO, this model synthesizes 'what is true in the medical and social models, without making the mistake each makes in reducing the whole, complex notion of disability to one of its aspects'.[91] Within the ICF, disability is used as 'an umbrella term for impairments, activity limitations and participation restrictions'.[92] Functioning, on the other hand, 'refers to body functions, activities and participation'.[93]

As can be seen from Fig. 4.1, the ICF divides the contextual factors which may contribute to a finding of disability, into environmental factors and personal factors. Environmental factors include social attitudes and architectural structures, whereas personal factors take into account 'gender, age, coping styles, social background, education… and other factors that influence how disability is experienced by the individual'.[94]

Alan Jette suggests that one of 'the most exciting aspects of the ICF framework is that it has the potential to provide a universal, standardized disablement language and framework that looks beyond mortality and disease to focus on how people live with their conditions'.[95] However, critics of the ICF framework argue that it is 'inadequate' as it 'explicitly relies on a medicalised understanding of disability… Therefore, disabled people are still defined as "problematic" because they are unable

---

90 World Health Organisation, *Towards a Common Language for Functioning, Disability and Health: ICF—The International Classification of Functioning, Disability and Health* (2002) 9.

91 Ibid.

92 Ibid 2.

93 Ibid.

94 Ibid 10.

95 Jette A (2006) Toward a common language for Function, Disability and Health. 86 *Physical Therapy* 726, 732.

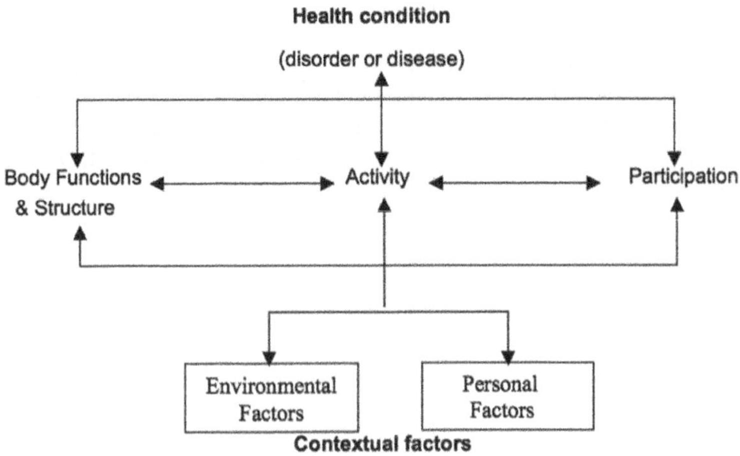

Fig. 4.1 The WHO's biopsychosocial model[96]

to conform to standards of normality'.[97] It has also been suggested that the ICF needs further refinement before it can be considered a complete model.[98] Gale Whiteneck asserts that the 'differentiation of activities and participation' requires further consideration, as the current bundling together of these factors is unsatisfactory.[99] He also points out that the model would benefit from the inclusion of quality of life,[100] 'elaboration of the impact of environmental factors [and] the development of personal factors'.[101] Further, Szilvia Geyh et al. contend that given the significance, relevance and usefulness of personal factors within the ICF, 'further standardisation' of these factors is required.[102] Despite criticism, there is general agreement that the

[96] Reprinted from World Health Organisation, *Towards a common Language for Functioning, Disability and Health: ICF—The International Classification of Functioning, Disability and Health* (2002) 9.

[97] Smith SR (2009) 'Social Justice and Disability: Competing Interpretations of the Medical and Social Models. In: Kristjana Kristiansen, Simo Vehmas and Tom Shakespeare (eds), *Arguing about Disability: Philosophical Perspectives* (Routledge) 15, 17.

[98] Whiteneck G (2005) Conceptual Models of Disability: past, present and future. In: Field M, Jette A, Martin L (eds), *Workshop on Disability in America: A New Look* (The National Academies Press) 50, 56.

[99] Ibid.

[100] Support for the inclusion of quality of life into the biopsychosocial model also comes from Sophie Mitra and Tom Shakespeare, who suggest that its addition would strengthen the model—Mitra S, Shakespeare T (2019) Remodeling the ICF. Disability and Health Journal 12: 337–339, 338.

[101] Whiteneck G (2005) Conceptual Models of Disability: past, present and future. In: Field M, Jette A, Martin L (eds) Workshop on Disability in America: A New Look. The National Academies Press, 50, 56.

[102] Geyh S et al (2011) The Personal Factors of the International Classification of Functioning, Disability and Health in the literature—a systematic review and content analysis. 33 *Disability and Rehabilitation* 1089, 1099.

ICF is preferential to its predecessor the International Classification of Impairments, Disabilities and Handicaps.[103]

4.2.6 The Political/Relational Model

The penultimate model of disability under consideration is Alison Kafer's political/relational model. Like the social model of disability, the political/relational model locates disability outside the human body.[104] Kafer refers to her model as a 'friendly departure' from the social model, and notes that the impairment–disability distinction does not form part of the political/relational model.[105] This is due to the social model's denial that impairment itself can ever be disabling, and its focus on disability as a social construct. Kafer, on the other hand, argues that '*both* impairment and disability are social', because 'social meanings' form part of our understanding of impairment.[106] As a consequence, under the political/relational model, the definition of impairment may shift 'across time and place, and presenting impairment as purely physical obscures the effects of such shifts'.[107]

Kafer's model moves away from the social model's rejection of medical intervention, instead treating medical interventions as neutral.[108] However more important for Kafer is 'social change and political transformation'.[109] Thus one of the aims of the political/relational model is to increase recognition of the 'political nature of a medical framing of disability'.[110] By acknowledging that disability is political, Kafer argues that her model paves the way for 'activist responses' to disability, thereby creating an opportunity for society to reimagine the concept of disability.[111]

One feature of the political/relational model is that it does not offer a firm definition of disability, instead seeing disability 'as a site of questions'.[112] Such questions assist in identifying the dimensions of disability, for example, does a person who is in remission from cancer have a disability? Is a person with a chronic illness disabled?[113] Kafer suggests that rather than relying on a fixed definition of

[103] McLean and Williamson, *Impairment and Disability: Law and Ethics at the Beginning and End of Life* (Routledge, 2007), 31; Hurst R, 'To Revise or not to Revise?' (2000) 15 *Disability & Society* 1083; Smith SR, 'Social Justice and Disability: Competing Interpretations of the Medical and Social Models' in Kristiansen K, Vehmas S, Shakespeare T (eds), *Arguing about Disability: Philosophical Perspectives* (Routledge, 2009) 15.

[104] Kafer A, *Feminist, Queer, Crip* (Indiana University Press, 2013) 6.

[105] Ibid 7.

[106] Ibid.

[107] Ibid.

[108] Ibid 6.

[109] Ibid.

[110] Ibid.

[111] Ibid 9.

[112] Ibid 11.

[113] Ibid.

disability, the framework within which disability is discussed should always be open for debate.[114]

4.2.7 A Feminist Material Account

The final approach to modelling disability considered here is Rosemarie Garland-Thomson's feminist material account of disability, which she refers to as 'misfits'.[115] Through material feminism, Garland-Thomson aims to shift the emphasis in feminist theory from operating at a level of discourse to focusing on the material—the relationship between the body and the built environment.[116] Garland-Thomson's account locates disability not in the body of the individual, nor in the social structures of their surroundings. Rather, the problem lies in the juxtaposition between them.[117] She likens this to trying to fit a square peg into a round hole—neither the peg nor the hole is inherently problematic, but attempting to fit them together gives rise to difficulty.[118] Like conceptions of disability within the social model and the interactional model, the concept of 'misfitting' changes as contexts shift.[119] Further, both misfitting and the interactional model emphasise the importance of the interaction between the person and their surroundings. The difference between the two models is that in advancing the concept of misfits Garland-Thomson does not explicitly refer to impairment.

Garland-Thomson asserts that the concept of misfitting is useful because it situates 'injustice and discrimination in the materiality of the world more than in social attitudes', while simultaneously recognising that the two are connected.[120] Therefore, 'misfitting serves to theorize disability as a way of being in an environment'.[121] When a body interacts with an environment capable of sustaining it in a harmonious way, a 'fit' has occurred. However, the built environment is often arranged in a way that is suitable for able-bodied people, and misfitting for people with disabilities.[122]

Garland-Thomson suggests that the notion of misfitting can extend beyond disability theory and have universal application 'as a contingent and fundamental fact of human embodiment'.[123] Misfits can thus form part of broader discussions on the uncertainty and instability of the body. This is important for Garland-Thomson because focusing 'on the contingency of embodiment avoids the abstraction of

---

[114] Ibid.

[115] Garland-Thomson R (2011) Misfits: A Feminist Material Disability Concept. Hypatia 26(3): 591–609.

[116] Ibid 594.

[117] Ibid 593.

[118] Ibid.

[119] Ibid.

[120] Ibid.

[121] Ibid 594.

[122] Ibid.

[123] Ibid 598.

persons into generic, autonomous subjects of individual liberalism'.[124] Therefore, an important part of the concept of misfitting is recognising that individuals are vulnerable, in the sense that everyone has the potential to misfit.[125]

4.2.8 A Suitable Model for Reproductive Choice?

Having considered various models of disability, it may be asked, what role a model of disability plays in the context of reproductive choice. For example, the medical and biopsychosocial models have clinical roles, such as the identification and treatment of disabilities, as well as administrative roles, including the allocation of resources such as welfare payments.[126] For the purposes of this book, a suitable model of disability must be capable of providing a theoretical basis for defining disability. This definition will be applied in the subsequent chapters of this book to inform discussions about how to develop a regulatory framework for embryo testing in Australia.

I propose the interactional model as a good starting point from which to make decisions regarding reproductive choice. This model combines the useful elements of both the medical and social models, thus providing a middle ground between the two traditional schools of thought. This choice would likely be supported by disability scholars such as Erik Parens, who refers to the medical and social models as 'lenses' through which we view disability, and argues that we should view disability through both these lenses; thus adopting a 'binocularity' approach to disability.[127] Further, conceptualising about disability as an interaction suggests that it is a flexible concept predicated upon an interface between two elements, rather than a static concept based on a single fixed point, for example, impairment.[128] One way in which the interactional model will be beneficial in the context of reproductive choices is that it may enable medical professionals and prospective parents to work together to decide whether or not a condition or disorder present in the embryo will actually be a disability. The medical professional is in a position to identify possible impairments for which preimplantation genetic testing (PGT) may be utilised, while the prospective parents are well placed to evaluate the social environment into which the child will be born and raised. Therefore, by considering the interaction between

---

[124] Ibid.

[125] Ibid.

[126] Altman BM (2001) Disability Definitions, Models, Classifications, Schemes and Applications. In: Albrecht GL, Seelman KD, Bury M (eds) Handbook of Disability Studies, SAGE, 97–122, 99–100.

[127] Parens E (2017) Choosing Flourishing: Toward a More "Binocular" way of Thinking about Disability. *Kennedy Institute of Ethics Journal* 27 (2):135–150.

[128] Support for the conceptualization of disability as an interaction is also found in the International Convention on the Rights of Persons with Disabilities. Article 1 of the Convention describes persons with disabilities as 'those who have long-term physical, mental, intellectual or sensory impairments which *in interaction with* various barriers may hinder their full and effective participation in society' (emphasis added).

these factors, a decision may be made regarding whether a particular condition may become a disability and consequently if testing should be undertaken. While the interactional model may be criticised for viewing impairment negatively, the benefit of this model is that in the context of reproductive choices, prospective parents may consider that the social factors are not sufficiently limiting and therefore that PGT is not required.

The welfarist account may also provide a useful insight for thinking about disability. The focus on individual wellbeing as a measure of disability could be a useful alternative to the traditional normal species functioning comparison for assessing disability. Therefore, in the context of reproductive choices, a helpful way to think about whether to test for a particular condition is to ask whether the future child's wellbeing would be higher if they did not have that condition. Similarly, there are aspects of the political/relational model that are desirable when thinking about disability in the context of reproductive choices. As reproductive decision-making involves decisions about future persons, it is beneficial to have a framework that is constantly evolving in light of changing political and economic conditions. Misfits also gives us a valuable perspective from which to consider disability in the sphere of reproductive choice. It reminds us that the 'body is dynamic',[129] and while a future child may experience times in their lives when their body and their environment will result in a misfit, prospective parents may be hopeful that fits will also occur. In a similar vein to the interactional model, these models all introduce a component of interaction to the conceptualisation of disability.

The interactional model is preferable to the WHO's biopsychosocial model because the latter was neither designed for, nor in its current form is it an appropriate framework for, reproductive decision-making. As the biopsychosocial model forms the basis of the ICF, its main applications are in the context of service provision.[130] Although the WHO states that the ICF may also be used in policy development, the types of policy uses envisaged appear to focus on individuals already living with disability, such as development of eligibility criteria for disability benefits and access to assistive technology.[131] This focus is further reinforced by the model's personal factors, many of which would be virtually impossible to apply to a person who is not yet born.

The common theme running through all the models of disability considered in this chapter is the association with limitations, reduced wellbeing and reduced capacities

---

[129] Garland-Thomson R (2011) Misfits: A Feminist Material Disability Concept. Hypatia 26(3): 591–609, 603.

[130] World Health Organisation, *Towards a Common Language for Functioning, Disability and Health: ICF—The International Classification of Functioning, Disability and Health* (2002), 6.

[131] Ibid 7.

for flourishing.[132] These limitations may be caused by social factors such as discrimination, as highlighted in the social model, by individual factors such as the loss of a limb, as discussed in the medical model, or by a combination of both; as exemplified in the interactional model. There is also general agreement that the social model's view of impairment as neutral does not reflect the reality of the situation for many people with disabilities. This consensus that disability is negative in nature suggests that individuals have an interest in avoiding disability. Further, this may also suggest that prospective parents utilising PGT should have the freedom to avoid disability in their offspring where possible.

4.3 Reproductive Liberty

Having outlined the theoretical framework for disability that will be drawn upon in the forthcoming chapters, this chapter now turns to theories of reproductive liberty. Setting out the theoretical underpinnings of the concept of reproductive liberty helps to answer the question which is central to the chapters which follow, namely, what kind of children should prospective parents have the freedom to create through the use of ART?

Reproductive liberty, sometimes referred to as procreative liberty, may be understood as the freedom to make decisions for oneself on matters of reproduction.[133] As the name suggests, reproductive liberty stems from the concept of liberty, as articulated by John Stuart Mill.[134] Mill argued that individuals should be free to follow their own life plan without interference, provided that plan does not harm others, 'even though they should think our conduct foolish, perverse or wrong'.[135] This is known as the harm principle and it has been suggested that this principle should be the 'threshold for justifying state intervention'.[136]

---

[132] There are, of course, other models and ways of thinking about disability that have not been covered in this chapter. For example, Desiree Valentine writes that disability should be understood as a technology, 'not just something that is produced but as a *tool* that organizes society and relations'. She notes that technologies are 'forces that reshape meaning and intentionality'. Thus, Valentine suggests that disability as a technology has the ability to reshape the discussion around reproductive choice. Valentine D (2021) Technologies of Reproduction: Race, Disability and Neoliberal Eugenics. Journal of Philosophy of Disability 1: 35–55.

[133] See generally, Dworkin R, *Life's Dominion: An Argument about Abortion, Euthanasia and Individual Freedom* (Vintage Books, 1994).

[134] Mill JS, *On Liberty* (Penguin Books, first published 1859, 1974 ed).

[135] Ibid 71.

[136] Smith MK, *Saviour Siblings and the Regulation of Assisted Reproductive Technology* (Ashgate, 2015) 86.

Reproductive liberty is closely related to the notion of reproductive autonomy.[137] While autonomy refers to a person's ability to make an informed decision,[138] liberty is concerned with establishing whether or not state intervention in certain areas of life is legitimate.[139] Therefore, reproductive liberty underpins reproductive autonomy because it gives individuals the opportunity to make reproductive decisions in line with their own values. Erin Nelson suggests that there are two aspects to reproductive autonomy—a personal dimension and a political one.[140] The personal dimension refers to the making of autonomous choices, whereas the political aspect pertains to the extent of state intervention. By framing reproductive autonomy in this way, Nelson encompasses reproductive liberty within her approach to autonomy.[141]

4.3.1 Procreative Liberty

One of the most influential and comprehensive approaches[142] to reproductive liberty is John Robertson's theory of procreative liberty. Procreative liberty comprises two aspects: 'the liberty to avoid having offspring and the liberty to have offspring'.[143] Robertson argues that 'Procreative liberty should enjoy presumptive primacy when conflicts about its exercise arise because control over whether one reproduces or not is central to personal identity, to dignity, and to the meaning of one's life'.[144] Robertson asserts that the freedom to have offspring encompasses the freedom to undertake activities or make decisions that culminate in the birth of a child.[145] As the liberty to have offspring is the relevant liberty for present purposes, it will be the focus of this section.

Robertson describes procreative liberty as a 'negative claim-right'.[146] He defines a 'claim-right' as a right that a person may exercise without violating a moral duty and one that others have a moral duty not to interfere with the exercise of. The negative aspect of the right refers to its nature as 'a right against interference by the

[137] Smith MK, *Saviour Siblings and the Regulation of Assisted Reproductive Technology* (Ashgate, 2015) 91.

[138] Ibid 92.

[139] Ibid.

[140] Nelson E, *Law, Policy and Reproductive Autonomy* (Hart, 2013) 36.

[141] See also Smith MK, *Saviour Siblings and the Regulation of Assisted Reproductive Technology* (Ashgate, 2015) 101.

[142] Roberts D (1997) Killing the Black Body. Pantheon Books, New York 296; Nelson E (2013) *Law, Policy and Reproductive Autonomy* (Hart, London) 32.

[143] Robertson JA, 'Procreative Liberty in the Era of Genomics' (2003) 29 *American Journal of Law and Medicine* 439, 447.

[144] Robertson JA (1994) Children of Choice. Princeton University Press, New Jersey 24.

[145] Robertson JA, 'Procreative Liberty in the Era of Genomics' (2003) 29 *American Journal of Law and Medicine* 439, 447.

[146] Robertson JA, 'Procreative Liberty in the Era of Genomics' (2003) 29 *American Journal of Law and Medicine* 439, 448.

state or others with reproductive decisions'.[147] In this way it may be distinguished from a positive right,[148] which may be framed in terms of a right to be provided with resources that enable a person to have a child; for example, state funded in vitro fertilisation (IVF). Consequently, 'the negative right to procreate ... does not imply the duty of others to provide the resources or services necessary to exercise one's procreative liberty despite plausible moral arguments for governmental assistance'.[149] Further, Robertson points out that procreative liberty is not an absolute right; however, 'there is a strong presumption in its favour, with the burden on opponents to show that there is a good case for limiting it'.[150] Therefore, Robertson suggests that the debate in this sphere concerns which exercises of procreative liberty might cause harm to others such that their exercise may be justifiably curtailed.[151] As procreative liberty enjoys presumptive primacy, 'substantial harm' would need to be caused to limit procreative choices, and as Robertson accepts, what counts as substantial harm will often be contentious.[152] This reflects Mill's harm principle. Therefore, the state should only interfere with reproductive decisions where those decisions pose a risk of harm to society or an individual, in this case the future child.

For Robertson, the liberty to reproduce is 'an important moral right' deserving of respect, due to the significance of reproduction to people's lives.[153] In order to decide whether limiting a person's procreative liberty is acceptable in a given situation, Robertson puts forward a two-stage test. The first question is whether the decision involves a 'distinctly procreative interest'.[154] If this is satisfied, the next question is whether the threat of harm from 'reproduction satisfies the strict standard for overriding this liberty interest'.[155] Robertson elaborates upon what would satisfy this 'strict standard' by suggesting that to 'have a liberty interest in procreating, one should at a minimum have the mental capacity to understand or appreciate the meanings associated with reproduction'.[156] Thus, according to Robertson, individuals with a severe intellectual disability who lack the capacity to appreciate the significance of reproduction do not hold the 'presumptive right' to procreative liberty.[157] On the

---

[147] Ibid.

[148] The idea of positive and negative rights stems from the concept of positive and negative liberty. A person can be said to have negative liberty in the areas of their life which are not subject to interference. Positive liberty, according to Isaiah Berlin, is concerned with being free to make decisions based on one's own goals and beliefs. Put another way, positive liberty is about having control over one's actions. See generally, Berlin I, edited by Henry Hardy, *Liberty: Incorporating Four Essays on Liberty* (Oxford University Press, 2002) 168–181.

[149] Robertson JA (1994) Children of Choice. Princeton University Press, New Jersey 23.

[150] Robertson JA, 'Procreative Liberty in the Era of Genomics' (2003) 29 *American Journal of Law and Medicine* 439, 448.

[151] Ibid.

[152] Robertson JA (1994) Children of Choice. Princeton University Press, New Jersey 24.

[153] Ibid 30.

[154] Ibid.

[155] Ibid.

[156] Ibid.

[157] Ibid 30–31.

other hand, unfit parents hold this right, even if they should be prevented from rearing the child once they are born.[158] Robertson suggests that the imposition of 'rearing costs on others may not rise to the level of harm that justifies depriving a person of a fundamental moral right'.[159]

It may be asked, which reproductive decisions fall within the scope of procreative liberty? Robertson asserts that activities that are 'so closely associated with, or essential to, reproductive decisions' would fall within the scope of procreative liberty.[160] Procreative liberty applies equally to individuals who are able to procreate naturally and those who cannot reproduce without assistance, as the moral right which it protects is the desire to have children.[161] Thus, the presumptive primacy of the right extends to the use of assisted reproductive technology (ART).[162] As well as the use of procedures such as in vitro fertilisation (IVF) that result in the creation of an embryo, Robertson suggests that the scope of procreative liberty may extend to actions such as acquiring genetic information about an embryo or foetus, or even about the mother herself, because such knowledge may determine whether the individual or couple decide to reproduce.[163] On the other hand, Robertson notes that activities that take place during the final stages of pregnancy, such as the decision to have a home-birth, would not fall under 'the protective canopy of reproductive liberty'.[164] This is because such decisions 'are not themselves determinative of whether reproduction will occur'.[165]

Embryo testing for the purpose of selecting against genetic conditions falls within the scope of procreative liberty, as noted above. However, Robertson questions whether 'intentional diminishment of offspring characteristics', as may be argued is the case with selection for disability, would fall within the scope of procreative liberty.[166] He suggests that such a use of embryo testing may 'deviate from the core interests that make reproduction meaningful as to fall outside the protective canopy of procreative liberty'.[167] The same quandary may arise in the context of sex selection for non-medical purposes, selection for traits that may be considered advantageous, and human leukocyte antigen typing (HLA typing). Whether reproductive

---

158 Ibid 31.

159 Ibid.

160 Robertson JA, 'Procreative Liberty in the Era of Genomics', (2003) 29 *American Journal of Law and Medicine* 439, 448.

161 Robertson JA (1994) Children of Choice. Princeton University Press, New Jersey 32.

162 Ibid.

163 Robertson JA, 'Procreative Liberty in the Era of Genomics', (2003) 29 *American Journal of Law and Medicine* 439, 448.

164 Ibid.

165 Ibid.

166 Ibid.

167 Robertson JA (1994) Children of Choice. Princeton University Press, New Jersey 34.

liberty extends to these uses of embryo testing will be discussed in the forthcoming chapters.

4.3.1.1 Critiquing Procreative Liberty

Robertson's view of procreative liberty as a negative claim-right has been seen as problematic as 'the negative view of liberty privileges the choices of those who have the means to realise them'.[168] As Dorothy Roberts explains, if 'reproductive freedom is important enough to human dignity to protect from government interference, then its infringement by forces other than the state should also be addressed'.[169] Further, she argues that liberal theorists, such as Robertson, have failed to give sound reasons 'why our understanding of procreative liberty must adopt a baseline of existing inequalities'.[170] Thus, for Roberts, a sound theory of reproductive rights needs to encompass a positive right to public services, so everyone can exercise their procreative liberty, not just those who can afford to do so. Similarly, Nelson notes that while Robertson's traditional liberal formulation of procreative liberty is preferable to an account that significantly constrains reproductive choices, 'it is ultimately unsatisfactory, as it will undoubtedly result in the denial of meaningful reproductive choice for those who are economically and socially disadvantaged'.[171] Robertson, on the other hand, suggests that whether the state should alleviate the social disadvantages that impact upon procreative liberty 'is a separate issue of social justice'.[172] Robertson's decision to omit a number of choices from the scope of procreative liberty, for example, a woman's choices during pregnancy,[173] have also been the subject of criticism.[174] Nelson contends that his 'exclusion of such aspects of reproduction from procreative liberty seems largely to discount the experience of reproduction from the reproducing woman's point of view'.[175]

While procreative liberty focuses on freedom from government interference, Robertson notes the importance of education in developing public policy, and the benefits of public education campaigns in areas pertaining to the avoidance of reproduction, such as the use of contraceptive devices or the termination of pregnancies.[176] However, no reference is made to education in the context of the freedom to procreate. Therefore, Robertson's conception of procreative liberty could benefit from a broader approach to education; one that covers both the freedom to procreate

[168] Roberts D (1997) Killing the Black Body. Pantheon Books, New York 296.

[169] Ibid.

[170] Ibid.

[171] Nelson E (2013) *Law, Policy and Reproductive Autonomy* Hart, London 34.

[172] Robertson JA (1994) Children of Choice. Princeton University Press, New Jersey 23.

[173] Robertson draws a distinction between 'liberty during the course of reproduction' and procreative liberty: ibid 23.

[174] Nelson E (2013) *Law, Policy and Reproductive Autonomy* Hart, London 33.

[175] Ibid.

[176] Robertson JA (1994) Children of Choice. Princeton University Press, New Jersey 47, 79–80.

and the freedom to avoid procreation. Public education in relation to the freedom to procreate can address issues like declining fertility, disability and the limits of ART. Thus, to enable individuals to fully exercise their reproductive liberty, they need to have knowledge of their choices and this is why public education is important.

Laura Purdy questions one of the key tenets of Robertson's model—the idea that procreation is central to a person's identity and dignity.[177] She argues that having a child does not a define a person and placing such high value on the ability to procreate could result in 'devastating disappointment' for some people.[178] The idea of reproduction as being of core importance to a person also places a disproportionate burden on women, who not only bear greater risks during reproduction, but may feel compelled to use ART, with its attendant financial, physical and emotional costs. Further, Purdy cautions against viewing the decision to reproduce as a primarily personal decision, as opposed to 'a moral decision affecting others'.[179] This is because the 'emphasis on self-creation' stifles any attempt to construct moral standards, due to a fear of violating a person's 'most intimate self'.[180]

4.3.2 Alternative Theories of Reproductive Rights

4.3.2.1 Reproductive Equality

An alternate theory of reproductive rights has been put forward by Radhika Rao. Rao argues that prospective parents do not have a 'general right to use ARTs as a matter of reproductive *autonomy*', but rather 'a limited right to use ARTs as a matter of reproductive *equality*'.[181] Rao's theory of reproductive equality requires that 'all persons must possess an *equal* right, even if no one retains an *absolute* right to use ARTs'.[182] Reproductive equality would effectively permit states to prohibit specific uses of reproductive technology, provided those uses are prohibited for all groups in society. Therefore, it would appear that prohibiting certain uses of embryo testing would be unproblematic provided all groups (for example, married and single people) are equally affected by the prohibition. However, Rao asserts that 'the government must set forth some legitimate interest in order to justify any selective prohibition upon the use of PG[T] to select for or against certain traits but not others'.[183] Rao

[177] Purdy L M (1996) Reproducing Persons: Issues in Feminist Bioethics. Cornell University Press, Ithaca 218.

[178] Ibid 219.

[179] Ibid.

[180] Ibid.

[181] Rao R, 'Equal Liberty: Assisted Reproductive Technology and Reproductive Equality' (2007) 76 *George Washington Law Review* 1457, 1460.

[182] Ibid.

[183] Ibid 1484.

does not elaborate on what would qualify as a legitimate interest, which leaves open the question whether reasons falling short of harm to the future child would suffice.

Rao suggests that reproductive equality is preferential to Robertson's procreative liberty because the latter has 'no logical stopping point'.[184] This is because in the US context in which both theories were conceived, procreative liberty appears to confer constitutional protection upon nearly every technology necessary for procreation.[185] As no such protection is provided by the Australian constitution,[186] at first glance Rao's reproductive equality appears to place a lower burden on the state to show that there are good reasons for limiting reproductive choices. Kimberly Mutcherson argues that Rao's account of reproductive equality is problematic because 'it does not take seriously the meaning of a right to procreate'.[187] For example, by drawing a distinction between those who can procreate naturally and those who need ART,[188] Rao 'relegates some to a life potentially without a fundamental right to procreate while preserving that right for others'.[189] Thus Mutcherson asserts that the equality for which Rao advocates 'is actually no equality at all'.[190]

4.3.2.2 The Procreative Right

An alternative to the above theories is Carter Dillard's theory of reproductive rights. Dillard argues that theories such as Robertson's construe reproductive liberty too broadly, giving rise to a broad procreative right which has no basis in law.[191] Dillard instead proposes a narrower right, not to procreate freely utilising any means desired, but instead, a right to procreate for the purposes of the continuation of humankind.[192] Dillard emphasises 'replacement and continuation' as the basis of the right.[193] A right to continue the species may extend to the use of ART, however, it does not provide any right to technologies such as embryo testing, apart from perhaps in cases where a chromosomal abnormality prevents a prospective mother from carrying a viable pregnancy. Further, the right may be limited 'by natural duties to prospective children, and, through them, duties to other members of society'.[194] Dillard notes that 'despite the common rhetoric, procreation is not protected by privacy or autonomy interests

[184] Ibid 1460.

[185] Ibid.

[186] *Commonwealth of Australia Constitution Act 1901* (Imp).

[187] Mutcherson K M (2015) Procreative Pluralism. Berkeley J. Gender L. & Just. 30: 22, 49.

[188] Rao R, 'Equal Liberty: Assisted Reproductive Technology and Reproductive Equality' (2007) 76 *George Washington Law Review* 1457, 1462–1465.

[189] Mutcherson K M (2015) Procreative Pluralism. Berkeley J. Gender L. & Just. 30: 22, 49.

[190] Ibid.

[191] Dillard C (2007) Rethinking the Procreative Right. 10 *Yale Human Rights and Development Law Journal* 1, 22.

[192] Ibid 38.

[193] Ibid 44.

[194] Ibid 43.

because it is not a personal or private act. Indeed, it is difficult to think of something less personal than creating another person'.[195]

Dillard distinguishes the right to procreate from the right not to procreate. He argues that the distinction turns on the right not to procreate involving only a woman and her body, whereas the right to procreate involves the creation of another human being.[196] Similarly, Rao argues that the distinction between the right to have a child and the right not to have a child centres on the difference between having one's bodily integrity respected through the removal or prevention of pregnancy on the one hand, and obtaining bodily interventions to become pregnant on the other.[197] Thus both Dillard and Rao ground the right to refrain from procreation in a right to privacy, whereas the right to procreate falls outside this sphere.

4.3.3 Choosing a Model of Reproductive Rights

The model of reproductive liberty adopted in this book is loosely based on Robertson's theory of procreative liberty. Robertson's theory has been selected because the conceptualisation of procreative liberty as a negative claim right with a broad scope opens the door to discussions about a range of uses of embryo testing. As the technology used to test embryos is constantly evolving, a model that requires more than 'theoretical or speculative' harm[198] in order to restrict reproductive choices is desirable. Although the negative claim right aspect of Robertson's work has been criticised, separating which uses of embryo testing are permissible for the purposes of developing a regulatory framework, from which uses the state should fund, aligns with the approach taken in this book. The loose adoption of procreative liberty allows me to dissociate certain aspects of Robertson's work from the model that will be applied in this book. For example, Robertson's classification of the right to procreate (or avoid procreation) as being central to the meaning of a person's life. Rather, the right to reproduce should be viewed as one aspect of a person's life. Grounding reproductive liberty in the freedom to makes decisions about various aspects of one's life does not in any way weaken the right. As I am not adopting Robertson's theory in its entirety, I shall refer to the model adopted here as reproductive liberty.

A model based on procreative liberty is preferable to one based on Rao's reproductive equality, because it limits state interference to cases where harm may arise, whereas Rao affords the state a potentially wider opportunity for interference. Procreative liberty is also more suitable than Dillard's model, because the latter presents a narrower reproductive right, the scope of which does not extend to the uses of embryo testing that are the subject of this book.

---

[195] Ibid 50.

[196] Ibid 51.

[197] Rao R (2007) Equal Liberty: Assisted Reproductive Technology and Reproductive Equality. 76 *George Washington Law Review* 1457, 1465.

[198] Robertson JA (1994) Children of Choice. Princeton University Press, New Jersey 35.

4.4 Conclusion

This chapter has introduced the concepts of disability and reproductive liberty. Disability is a complex concept as the range of conditions it encompasses varies significantly. It is nevertheless necessary to understand what falls within the notion of disability in order to advance discussions on reproductive choice. To ensure that disability is understood as encompassing genetic conditions such as cystic fibrosis and Huntington's disease (which are often considered as illnesses rather than disabilities), the term 'disabling condition' will be used in the subsequent chapters of this book. The model of disability that will be applied in the following chapters is the interactional model. This model considers disability as an interaction between the impairments a person may have and the environment into which they will be born. The benefits of this model include the flexibility inherent in the idea of an interaction and the corresponding advantages this presents in conversations about reproductive decision-making.

The concept of reproductive liberty was then outlined to provide a theoretical framework for discussions around parental choice. Reproductive liberty is the freedom to makes one's own reproductive choices. The model of reproductive liberty that will be adopted in the chapters that follow is based on Robertson's theory of procreative liberty. The conception of procreative liberty as a 'negative claim-right' requires that prospective parents' decisions in matters of procreation be respected unless significant harm to an individual or society will result. The scope of procreative liberty extends to the use of ART because one of the features of this model is that it applies equally to the fertile and those who are infertile or unable to reproduce without assistance. It should be noted that one aspect of Robertson's theory will not be applied in this book—the freedom to procreate will not be viewed as central to the meaning and dignity of one's life, but instead as one of many valuable life experiences which one may choose to undertake.

This chapter also considered other theories of reproductive rights, including Rao's reproductive equality and Dillard's theory of the procreative right to reproduce for the purposes of continuing the species. These theories are more restricted in their scope than Robertson's, and while worthy of discussion to understand the varying perspectives on reproductive rights, they will not play a significant role moving forward. Ultimately, the theories of disability and reproductive liberty advanced in this chapter will be instrumental in answering the question, what kind of children should prospective parents have the freedom to create through ART? In the next chapter, I will consider this question in relation to the use of PGT to test for disabling conditions.

Chapter 5
Selecting Against Disabling Conditions

Abstract Chapter 5 focuses on the ethical, legal, and social dimensions of testing embryos for disabling conditions. Building on the interactional model of disability introduced in the previous chapter, which frames disability as a product of both individual impairments and societal structures, this chapter evaluates whether prospective parents should have the liberty to test for and select against disabling genetic conditions. It begins by addressing objections to this use of PGT, starting with the expressivist argument, which claims that selecting against disabling conditions conveys negative messages about the worth of people living with such conditions. The chapter challenges this view by distinguishing between reducing the incidence of disabilities and devaluing individuals with disabilities. Further, the chapter examines arguments centring around eugenics, the loss of support for those with disabling conditions and the principle of procreative beneficence. The chapter advocates for a revised regulatory framework that allows parents to select against any conditions defined as disabling, thus enabling informed reproductive choices and safeguarding parental autonomy.

5.1 Introduction

The preceding chapter set out the theoretical basis of the concept of disability and introduced the interactional model of disability. This model defines disability 'as an interaction between individual and structural factors'.[1] Thus disability, in the context of reproductive choice, is the interaction between the impairments a future child may have due to a genetic condition and the social structures present in their surrounding environment. Following on from chaps. 2 and 3, which provided an overview of the existing regulatory frameworks for embryo testing and considered how they could be improved, this chapter and the two that follow examine what the content of a revised framework should be. In other words, what kind of children should prospective parents have the freedom to create using assisted reproductive technology (ART)?

---

[1] Shakespeare T (2006) *Disability Rights and Wrongs* (Routledge) 55.

M. de Souza, *The Regulation of Embryo Testing in Australia*,
https://doi.org/10.1007/978-981-96-5902-9_5

This Chapter discusses whether prospective parents should have the freedom to test for and select against disabling conditions.[2] In the previous chapter, the scope of reproductive liberty was said to extend to acquiring genetic information about an embryo, as the results of such tests may determine whether a person chooses to reproduce.[3] While many consider this to be an acceptable use of preimplantation genetic testing (PGT), disability rights advocates argue, among other things, that the use of PGT for this purpose expresses negative views about people with disabling conditions and is a eugenic practice. Thus, it is important to consider the weight of these arguments in the development of a revised framework for embryo testing. Following on from this, I look at the current framework which permits testing for disabling conditions, albeit with the restriction in some states that testing be confined to serious disabling conditions. In developing a national framework for the regulation of embryo testing, it is necessary to consider how this discrepancy between states should be resolved. Finally, the chapter ends with the argument that parents should be permitted to test for and select against any conditions that may be defined as disabling in order to give their child an open future.

5.2 *The Use of PGT to Select Against Disabling Conditions*

In the previous chapter, prospective parents' reproductive liberty was said to extend to the use of PGT to select against disabling conditions in their future children. This may include PGT for monogenic conditions (PGT-M), PGT for structural relocations (PGT-SR), or PGT for aneuploidy (PGT-A).[4] As being able to have a child without a specific genetic condition may determine whether or not a person chooses to reproduce, the use of PGT falls within the scope of reproductive liberty.[5] This section begins by considering the arguments advanced against the use of PGT to select against disabling conditions. Do any of these arguments justify limiting the reproductive liberty of prospective parents? First, I analyse the most common objection against the use of PGT for this purpose; the expressivist objection. Secondly, I explore the loss of support objection. I then ask whether selecting against disabling conditions is a eugenic practice and whether this is a reason not to undertake such

[2] As noted in Chap. 4, I will use the term disabling condition as some genetic conditions for which preimplantation genetic testing is undertaken are often considered to be illnesses rather than disabilities (e.g. cystic fibrosis). This term has been used by others such as Scully JL (2020) Disability, Disablism, and COVID-19 Pandemic Triage. Journal of Bioethical Inquiry 17: 601–605; Freeman L, Delatycki MB, Scully JL, Kirk EP (2022) Views of Reproductive Genetic Carrier Screening Participants Regarding Screening for Genes Associated with Non-syndromic Hearing Loss. Prenat Diagn. 42(13): 1658–1666.

[3] Robertson JA, 'Procreative Liberty in the Era of Genomics', (2003) *American Journal of Law and Medicine* 29: 439, 448.

[4] The abbreviation PGT will be used when discussing conceptual arguments that apply to all three types of PGT.

[5] Robertson JA (1994) Children of Choice. Princeton University Press, New Jersey, 33.

testing. Having shown that none of these arguments present a good enough reason to prevent or even discourage parents from using PGT to select against disabling conditions, I turn to the reasons put forward in favour of using PGT in this manner. Within this discussion, the principle of procreative beneficence and the open future argument will be will examined. Lastly, I consider late onset and lower penetrance conditions and ask whether PGT-M should be permitted for genetic conditions that only manifest themselves later in life.

5.2.1 The Expressivist Objection

The expressivist objection to PGT is the argument that testing for and selecting against disabling conditions reflects negative or discriminatory attitudes towards people living with such conditions.[6] The argument goes, by choosing to implant only embryos that will not develop into a child with a disabling condition, prospective parents are implying that the lives of people with disabilities are not worth living and that it would have been better if they had not been born. An alternative version of this argument is that by permitting testing of this kind, society is sending a negative message to individuals with disabilities.[7] Proponents of the expressivist objection argue that by not implanting an embryo on the basis that it will become a child with a disability, the value of the future child is being determined by a single characteristic.[8] As Erik Parens and Adrienne Asch note, when prospective parents make such a decision they let a 'part stand in for the whole'.[9] This is considered problematic because it reduces the worth of a future child to whether or not they have a genetic condition.[10]

The expressivist objection to PGT has been widely criticised for a number of reasons. First, the argument does not distinguish between disabling conditions and persons with disabling conditions.[11] For example, trying to avoid accidents which may render one paraplegic does not mean that people with paraplegia do not deserve

[6] The expressivist objection was originally advanced in the context of prenatal diagnosis. See, eg, Erik Parens and Adrienne Asch, 'The Disability Rights Critique of Prenatal Genetic Testing' (1999) 29(5) *Hastings Center Report* S1, S2.

[7] Holm S, 'The Expressivist Objection to Prenatal Diagnosis: Can it be laid to rest?' (2008) 34 *Journal of Medical Ethics* 24, 24.

[8] Parens E, Asch A, 'The Disability Rights Critique of Prenatal Genetic Testing' (1999) 29(5) *Hastings Center Report* S1, S2.

[9] Parens E, Asch A, 'The Disability Rights Critique of Prenatal Genetic Testing' (1999) 29(5) *Hastings Center Report* S1, S2.

[10] Malek J, 'Deciding Against Disability: Does the use of reproductive genetic technologies express disvalue for people with disabilities?' (2010) 36 *Journal of Medical Ethics* 217, 217.

[11] Savulescu J, 'Procreative Beneficence: Why we should select the best children' (2001) 15 *Bioethics* 413, 423.

respect.[12] Janet Malek illustrates this point well in a series of thought experiments, the first of which is below:

> you are walking down the street one sunny morning to your local coffee shop. The walk signal flashes and you are about to enter the intersection when everything around you suddenly freezes. A disembodied voice claiming to be a guardian angel informs you that a driver approaching the intersection will momentarily be blinded by the early morning sun and run the red light; you will be hit by the driver in the crosswalk and paralysed from the waist down. Time then restarts and you are faced with the choice of when to cross the street. What would you do? Assuming that you can get on board with the idea that that you have a guardian angel with the ability to freeze time, it seems safe to say that you would stay put on the sidewalk.[13]

Malek notes that by not stepping into the road we are selecting 'one possible future over another'.[14] However, she points out that this does not send a negative message to people who have paraplegia. Further, she states that

> Just as trying to prevent the occurrence of disability in your own life does not imply that you would find your own life less valuable if you were to become disabled, seeking to prevent disability in your own life does not imply that the lives of others with disability are any less valuable.[15]

Malek's second thought experiment posits the same situation, but this time you have a nine year old child and should the accident occur it will be the child who will become paraplegic.[16] Once again, the logical response would be to avoid the accident and Malek notes that this choice similarly does not indicate any disvalue for people with paraplegia, it simply 'suggests that having the ability to use one's legs is, all else being equal, a good thing.'[17] Malek's penultimate thought experiment is the most relevant for our purposes:

> You have just come from your local coffee shop and have arrived at your appointment with a genetic counsellor. You and your partner are thinking about trying to conceive a child but are aware that you have a family history of a serious genetic condition that causes those who inherit a particular gene to become paraplegic. The counsellor informs you that it is possible to use PGD to ensure that your future child will not have this condition. If you choose to take this approach, several eggs will be fertilised in vitro and allowed to grow into embryos. Those embryos will be tested for the genetic condition in question, and only those embryos that do not carry that gene will be transferred and given the chance to develop. You and your partner decide to use PGD to ensure that your future child will not have the genetic condition.[18]

Malek concludes that by choosing to avoid disability in their offspring, prospective parents are not devaluing the lives of people with disabling conditions because the

---

[12] Ibid.

[13] Malek J, 'Deciding Against Disability: Does the use of reproductive genetic technologies express disvalue for people with disabilities?' (2010) 36 *Journal of Medical Ethics* 217, 218.

[14] Ibid.

[15] Ibid.

[16] Ibid. 219.

[17] Ibid.

[18] Ibid 220.

use of PGT 'to prevent disability in a future child is no different from the decision to protect oneself from becoming disabled with regard to the message that such a choice sends to those individuals who currently have disabilities'.[19]

Turning to the alternative version of the expressivist objection, Malek argues that public education campaigns encouraging drivers to wear seatbelts, parents to vaccinate their children or pregnant mothers to refrain from smoking and drinking, do not express disvalue for individuals with disabling conditions.[20] Using the same line of reasoning that applies to individual choices, these public campaigns are suggesting that it would be better if children were born healthy and drivers did not sustain serious injuries, however, they are not implying that the lives of people with serious injuries or congenital birth defects are not worth living. Therefore, by distinguishing between disabling conditions and individuals with disabling conditions, society is able to provide prospective parents with the freedom to use PGT to select against disabling conditions, while at the same time respecting the lives of people living with such conditions.

Not everyone agrees that the lives of individuals with disabling conditions can be completely disentangled from their disabling conditions. Steven Edwards argues that some individuals with disabling conditions view their condition as part of their identity and it is for this reason that selection against disability is hurtful to these individuals.[21] Edwards refers to this as the identity claim.[22] However, even after accepting the identity claim, the expressivist objection is still problematic. If selecting against a disabling condition through PGT reduces the incidence of that disabling condition and is thus hurtful to individuals who consider that disabling condition to be part of their identity, it follows that attempts to cure or remedy the condition would also be hurtful to people with that condition.[23] This is because both PGT and medical intervention ultimately have the effect of reducing the incidence of the disability in question. As it would be ridiculous to argue that people should be prohibited from undergoing treatment to cure a disability, it would be equally to absurd to suggest that people should be prevented from using PGT to select against disability in their future offspring. Conversely, disability activists argue that medical treatment and PGT differ in one important way; medical treatment eliminates the disability while leaving the person intact, however, PGT eliminates the disability by effectively eliminating the future person.[24] As Colin Gavaghan asserts, 'it is not

---

[19] Ibid 221.

[20] Ibid 219–220.

[21] Edwards SD, 'Disability, Identity and the Expressivist Objection' (2004) 30 *Journal of Medical Ethics* 418, 418.

[22] Ibid.

[23] Ibid 419.

[24] Parens E, Asch A, 'The Disability Rights Critique of Prenatal Genetic Testing' (1999) 29(5) *Hastings Center Report* S1, S3.

simply the disorder that is being rejected, but the whole package of person-plus-disorder'.[25] Gavaghan suggests that some medical treatments, particularly those for mental illnesses and Alzheimer's disease, also have the effect of changing a person's identity, thereby effectively replacing the ill person with a different one.[26] Therefore, if the treatment of individuals with mental illness or cognitive impairments is to be permitted and even encouraged, the same attitude should apply to the use of PGT by prospective parents.

A further reason the expressivist objection is not a persuasive reason to stop prospective parents from using PGT to select against disability is that the hurt caused to individuals with disabling conditions is not a sufficient justification to curtail reproductive choices.[27] Timothy Murphy notes that the hurt caused by prospective parents' use of PGT does not reach a high enough threshold to justify state intervention prohibiting such testing.[28] Moreover, he points out that unless the effect of PGT is to cause direct harm to the wellbeing of people with disabling conditions, 'or indirect harm of a magnitude equivalent to direct harm, these choices remain morally defensible'.[29] Therefore, as long as the use of PGT does not lead to discrimination against or violence towards individuals with disabling conditions, there appears to be no reason to prohibit selection against disabling conditions.

The use of PGT to select against disability may inadvertently send hurtful messages to individuals living with disabling conditions; however, merely causing hurt to others is not an adequate justification for the restriction of reproductive liberty. As Parens and Asch note, prospective parents are not seeking 'to hurt existing disabled people but to implement their own familial goals. In that sense, there is no "message" being sent at all'.[30] Similarly, Shakespeare suggests that there may be other reasons, such as financial concerns (e.g. one parent having to give up work to care for the child), that lead prospective parents to use PGT.[31] Nevertheless, to minimise the hurt caused, society should aim to send positive messages to people with a disability that they are valued and respected members of society.[32] One way this could be done is through the provision of adequate care and assistance for individuals living with disabling conditions.

---

[25] Gavaghan C, 'Right Problem, Wrong Solution: A Pro-Choice Response to "Expressivist" Concerns about Preimplantation Genetic Diagnosis' (2006) 16 *Cambridge Quarterly of Healthcare Ethics* 20, 26.

[26] Ibid 27.

[27] Murphy TF, 'When Choosing the Traits of Children is hurtful to others' (2011) 37 *Journal of Medical Ethics* 105.

[28] Ibid 108.

[29] Ibid.

[30] Parens E, Asch A, 'The Disability Rights Critique of Prenatal Genetic Testing' (1999) 29(5) *Hastings Center Report* S1, S3.

[31] Shakespeare T (2014) Disability Rights and Wrongs Revisited. Routledge, Abingdon, 128.

[32] See generally, Nelson JL, 'Hurts, Insults and Stigmas: a comment on Murphy' (2011) 37 *Journal of Medical Ethics* 66; Glover J (2006) *Choosing Children: Genes, Disability and Design* (Oxford University Press), 35.

5.2.2 The Loss of Support Objection

The loss of support objection[33] to PGT is the argument that selection against disabling conditions will lead to less tolerance and consequently less public funded support for individuals with such conditions.[34] The objection is based on the idea that widespread use of PGT will reduce the number of people who are born with disabling conditions, thus reducing the impetus to find cures or treatments for particular conditions, or to make society more accessible.[35] Gavaghan notes that it 'is not implausible' that there would, for example, be less of an incentive to make buildings accessible if there are much smaller numbers of people using wheelchairs.[36] However, Tom Shakespeare points out that even if the use of PGT becomes widespread, it is unlikely that all prospective parents will use such testing, thus reducing, but not eliminating, the number of people born with disabling conditions.[37] Further, even if the use of PGT became widespread and the incidence of certain disabling conditions was reduced, the prevalence of much of what is considered to fall within the broad concept of disability (e.g. using a wheelchair) would be unaffected. This is because most disabilities are the result of factors other than genetics, such as ageing, accidents and war.[38]

As with the expressivist objection, there are a number of problems with the loss of support objection. First, falling numbers of people with disabling conditions could just as easily occur through advances in medical treatment and improved occupational health and safety laws, as through PGT.[39] Therefore, if reducing the number of individuals with disabling conditions through the use if PGT is objectionable, the loss of support objection appears to be suggesting that individuals should be denied medical treatment and perhaps even that workplace health and safety laws should be abolished. These conclusions do not appear sensible and provide just one example of why the loss of support objection is not a good reason to curtail parental choices.

Second, the loss of support objection is based on the assumption that a decrease in the prevalence of genetic conditions will lead to a decrease in support for individuals

[33] This expression has been adapted from Allen Buchanan's article where he terms it the 'loss of support argument'. Buchanan A, 'Choosing Who will be Disabled: Genetic Intervention and the Morality of Inclusion' (1996) 13(2) *Social Policy and Philosophy* 18, 21.

[34] Ibid.

[35] Gavaghan C, *Defending the Genetic Supermarket: Law and Ethics of Selecting the Next Generation* (Routledge Cavendish, 2007) 104.

[36] Gavaghan C, *Defending the Genetic Supermarket: Law and Ethics of Selecting the Next Generation* (Routledge Cavendish, 2007) 105.

[37] Shakespeare T (1995) Back to the Future: New Genetics and disabled people. Critical Social Policy, 15: 22–35, 31.

[38] Scully JL (2008) Disability and Genetics in the Era of Genomic Medicine. Nature Reviews Genetics 9: 797–802, 799. https://doi.org/10.1038/nrg2453.

[39] Buchanan A, 'Choosing Who will be Disabled: Genetic Intervention and the Morality of Inclusion' (1996) 13(2) *Social Policy and Philosophy* 18, 23; Gavaghan C, *Defending the Genetic Supermarket: Law and Ethics of Selecting the Next Generation* (Routledge Cavendish, 2007) 106.

with disabling conditions. For this argument to be persuasive, more than a generalisation is needed, 'Data to support it must be supplied'.[40] In the absence of empirical data showing the correlation between the prevalence of disability and support for individuals with disabling conditions, the objection lacks the weight required to interfere with reproductive liberty. In the context of prenatal diagnosis, which also reduces the incidence of people born with disabling conditions, Shakespeare asserts that such procedures have not resulted in reduced quality of life for those living with disabilities. He writes that 'the increasing availability of genetic knowledge has coincided with the increasing acceptance of disability rights and slowly improving provision for disabled people'.[41] Further, the loss of support objection appears counterintuitive because a limited number of resources will stretch further if there was less demand for those resources.

The case of orphan drugs is one example of incidence of a condition not being linked to the level of support available. Orphan drugs are drugs that are used to treat rare diseases. These drugs are expensive to manufacture and only benefit small numbers of people. While orphan drugs are unlikely to ever become cost effective, in the United Kingdom there have been suggestions that the National Health Service should, in certain circumstances, pay premium prices for these drugs.[42] Moreover, a number of jurisdictions, including Australia,[43] give special status to orphan drugs, as there is agreement that the development of these drugs is in the public interest and to deny treatment to patients suffering from rare diseases would be inequitable.[44] The case of orphan drugs may provide an indication of what would transpire if fewer people with disabling conditions were born. It is also an example of the prevalence of disabling conditions not being directly proportional to the level of support available.

Third, the loss of support objection is overly simplistic, as there are a number of factors that will influence the level of support given to individuals with disabling

---

[40] Buchanan A, 'Choosing Who will be Disabled: Genetic Intervention and the Morality of Inclusion' (1996) 13(2) *Social Policy and Philosophy* 18, 21.

[41] Shakespeare T (2014) Disability Rights and Wrongs Revisited. Routledge, Abingdon, 127.

[42] National Institute of Clinical Excellence, *NHS should consider paying premium prices for drugs to treat patients with very rare diseases says NICE Citizen's Council* (26 January 2005) https://www.ncbi.nlm.nih.gov/books/NBK401721/pdf/Bookshelf_NBK401721.pdf. The NICE Citizen's Council suggested that in deciding whether to fund treatment the three main criteria should be the seriousness of the disease, if the treatment will actually improve, as opposed to stabilize, the condition and whether or not the condition is life threatening. See also Department of Health and Social Care (UK), UK Rare Disease Framework (2021).

[43] In Australia, the Orphan Drugs Program is administered by the Therapeutics Goods Administration and aims to encourage the introduction of orphan drugs into the Australian market by providing financial incentives such as fee waivers. Department of Health and Aged Care, Australian Government (2018) Orphan Drug Designation https://www.tga.gov.au/resources/resource/guidance/orphan-drug-designation. Accessed 28 February 2023.

[44] McCabe C, Claxton K, Tsuchiya A (2005) 'Orphan Drugs and the NHS: should we value rarity?' 331 *British Medical Journal* 1016, 1016. See also *Regulation (EC) No 141/2000 of the European Parliament and of the Council of 16 December 1999 on orphan medicinal products* [2000] OJ L 18/1.

conditions, not merely their numbers.[45] Allen Buchanan suggests that if the public are well informed about the dangers associated with reduced support, this will have a positive impact on levels of social support.[46] He also points to the important role played by disability rights advocates in raising awareness of the importance of support.

The most compelling argument against the loss of support objection is that even if selection against disability did turn out to have a negative impact on the level of support given to individuals with disabilities, this does not make the practice 'wrong or even inadvisable'.[47] As Buchanan states:

> The problem with the "loss of support" argument is that it only considers the interests of those who will have disabilities after a reduction in the incidence of disabilities occurs. It entirely overlooks the interests that others have in *not having* disabilities.[48]

The loss of support objection also overlooks the interests parents have in maintaining their reproductive liberty. It would thus seem strange to require parents to bring up a child with a disabling condition purely for the sake of individuals already living with such conditions. Having uncovered a number of flaws with the loss of support objection, I suggest that this objection does not adequately justify curtailing the reproductive liberty of prospective parents.

5.2.3 Eugenics

Eugenics is a term that often appears in the context of PGT and generally has negative connotations. This is mainly due to the term's association with the atrocities that took place in Nazi Germany, including forced sterilisation and genocide of those considered unworthy to be part of Hitler's master race. Eugenic ideas can, however, be traced back to the ancient Greeks.[49] For example, eugenic themes are found in Plato's Republic where he writes about creating a city with 'the finest possible progeny'.[50] Plato proposed to achieve this through the prudent pairing of those with desired attributes.[51]

In the early 1900s, statutes permitting involuntary sterilisation were enacted in the name of eugenics in countries such as the United States and Denmark. The world's

---

[45] Buchanan A, 'Choosing Who will be Disabled: Genetic Intervention and the Morality of Inclusion' (1996) 13(2) *Social Policy and Philosophy* 18, 22.

[46] Ibid.

[47] Ibid.

[48] Ibid.

[49] Galton DJ (1998) Greek Theories on Eugenics. Journal of Medical Ethics 24: 263–267.

[50] Ibid 263.

[51] Ibid 264.

first 'modern eugenic sterilization law' was passed in Indiana in 1907.[52] The aim of the *1907 Indiana Eugenics Law* was to prevent certain groups from procreating, namely, 'confirmed criminals, idiots, imbeciles and rapists'.[53] The 1907 Act required institutions to convene a board of medical experts whose role was to make decisions as to whether or not it was desirable for such persons to procreate.[54] In Europe, Denmark was the first country to enact sterilisation legislation, on 1 June 1929.[55] Unlike the Indiana statute, the word eugenics did not appear in the title of the Danish Act, nor in its text.[56] The *Sterilization Act of 1929* (Denmark) had two main aims, one concerned with public security and the other with reproduction.[57] Thus s 1 of the statute provided for the castration of sex offenders,[58] while s 2 permitted the sterilisation of those with an intellectually disability. The Act considered it to be in the best interests of both society and individuals with an intellectual disability that they be prevented from having children.[59]

This aside, what does the word 'eugenics' mean? Is the use of PGT to select against disabling conditions a eugenic practice and if so, is this a good enough reason to curtail the reproductive choices of prospective parents? The word eugenics comes from the Greek words '*eu*' and '*gen*', which together roughly translate to good birth. In 1904, Francis Galton defined eugenics as 'the science that deals with all influences that improve the inborn qualities of a race; also with those that develop them to the utmost advantage'.[60] The Macquarie Dictionary defines eugenics as 'the theory or practice of improving the genetic characteristics of a population, especially

[52] Hansen BS, 'Something rotten in the state of Denmark: Eugenics and the ascent of the welfare state' in Broberg G, Roll-Hansen N (eds), *Eugenics and the Welfare State: Sterilization Policy in Denmark, Sweden, Norway and Finland* (Michigan State University Press, 1996) 9, 14.

[53] *1907 Indiana Eugenics Law*, Ind Code ch 215.

[54] Ibid.

[55] Hansen BS, 'Something rotten in the state of Denmark: Eugenics and the ascent of the welfare state' in Broberg G, Roll-Hansen N (eds), *Eugenics and the Welfare State: Sterilization Policy in Denmark, Sweden, Norway and Finland* (Michigan State University Press, 1996) 9, 10. It has been suggested that the first European eugenic statute may have been enacted by a Swiss state earlier in 1929—Wildenskov HO (1932) Sterilization in Denmark: A Eugenic as well as a Therapeutic Clause. Eugenics Review 23(4): 311–313, 311.

[56] Hansen BS, 'Something rotten in the state of Denmark: Eugenics and the ascent of the welfare state' in Broberg G, Roll-Hansen N (eds), *Eugenics and the Welfare State: Sterilization Policy in Denmark, Sweden, Norway and Finland* (Michigan State University Press, 1996) 9, 38.

[57] Wildenskov HO (1932) Sterilization in Denmark: A Eugenic as well as a Therapeutic Clause. Eugenics Review 23(4): 311–313, 311.

[58] The Act did not draw a distinction between violent sex offenders and others, such as homosexuals, who at the time were also considered to fall into the sex offender category.

[59] For a detailed history of eugenics in Denmark, see Hansen BS, 'Something rotten in the state of Denmark: Eugenics and the ascent of the welfare state' in Broberg G, Roll-Hansen N (eds), *Eugenics and the Welfare State: Sterilization Policy in Denmark, Sweden, Norway and Finland* (Michigan State University Press, 1996) 9–76. For a comprehensive history of the eugenics movement in other countries, see Kevles DJ (1998) In the Name of Eugenics: Genetics and the Uses of Human Heredity. Harvard University Press, Cambridge.

[60] Galton F (1904) Eugenics: Its Definition, Scope and Aim. American Journal of Sociology 10(1): 1–25.

by controlling reproduction in order to produce offspring with qualities considered desirable'.[61] A similar definition can be found in the Routledge Encyclopedia of Philosophy, where eugenics is defined as 'the attempt to improve the human gene pool'.[62]

What these definitions share is the idea that eugenics is concerned with improving future generations. The modern definitions (in the Macquarie Dictionary and the Routledge Encyclopedia of Philosophy) also imply an intention to do so; however, whether to qualify as eugenic a practice must be intentional is controversial.[63] Stephen Wilkinson and Eve Garrard note that unintentional eugenics would be difficult to distinguish from evolutionary processes and it would be unfortunate to have to refer to evolution as eugenic.[64] Further, they suggest that 'some people want "eugenics" to be a moral term, to have wrongness built into it, so that to call something "eugenics" is inevitably to condemn it'.[65] However, they assert that a 'morally neutral definition' is desirable for it enables discussion about whether any eugenic practices are appropriate.[66] Constructing a morally neutral definition of eugenics is difficult, because, as noted above, definitions of eugenics are based on the idea of improvement, and defining improvement, particularly in the context of offspring, may require a moral positioning that is not neutral. For example, Wilkinson himself writes that to refer to something as an improvement is to suggest that it is 'a change for the better'[67]; but the word 'better' means different things to different people.

Taking the above definitions as a starting point, eugenics may be said to include practices that aim to improve the human gene pool. Therefore, it needs to be asked if PGT is a eugenic practice. It is suggested here that PGT is not a eugenic practice. Selecting against disabling conditions, which as discussed in the previous chapter are associated with limitations,[68] may inadvertently improve the gene pool. For example, PGT-M can be used to select against conditions such as Huntington's disease. Huntington's disease is a serious, neurodegenerative condition that significantly affects the lives of people living with it.[69] As Huntington's disease is an autosomal dominant condition, if one prospective parent has the gene, there is a 50% chance a child will not only inherit the condition, but also pass it on to the next generation. If future generations are born without the Huntington's gene, this could be considered as an

[61] *Macquarie Dictionary* (Macquarie, 7th ed, 2017).

[62] Chadwick R, 'Genetics and Ethics' in Edward Craig (ed), *Routledge Encyclopedia of Philosophy* (Routledge, 1998).

[63] Wilkinson S, Garrard E, 'Eugenics and the Ethics of Selective Reproduction' (Report, Keele University, 2013) 2, n 4.

[64] Ibid.

[65] Ibid 3.

[66] Ibid.

[67] Wilkinson S (2010) Choosing Tomorrow's Children: The Ethics of Selective Reproduction. Oxford University Press, Oxford, 160.

[68] As discussed in Chap. 4, these limitations may be a result of the condition itself, a consequence of societal barriers, or due to a combination of both.

[69] Van Rij MC et al. (2012) Preimplantation Genetic Diagnosis (PGD) for Huntington's Disease: The Experience of Three European Centres. *European Journal of Human Genetics* 20: 368, 368.

improvement to the gene pool. However, it is not an intentional improvement. Intention implies that an action is undertaken with the aim of achieving a certain outcome (e.g. improving the gene pool). Prospective parents are acting in the best interests of their future child, without any concern for the human gene pool more broadly.[70] Therefore, giving parents the freedom to select against disability in their offspring does not fall within the meaning of eugenics. As a result, there appears no sound justification for curtailing reproductive choices in this area, purely on the basis that such choices are eugenic.

Not everyone agrees that the use of PGT is not a eugenic practice. For example, Jonathan Anomaly suggests that 'Eugenics can be thought of as any attempt to harness the power of reproduction to produce people with traits that enable them to thrive'.[71] Anomaly does not refer to improving the gene pool and thus his definition differs from the ones considered above. In the previous chapter, the interactional model was put forward as a starting point for reproductive decisions in the context PGT. According to this model, a disabling condition is one in which an individual's impairments interact with their environment in such a way as to cause functional limitations. Therefore, according to Anomaly's definition of eugenics, PGT-M would be considered a eugenic practice, because if a person is born without a condition that will cause them functional limitations they are more likely to thrive. However, not everyone agrees with Anomaly's definition of eugenics.[72] Robert Wilson argues that Anomaly has mischaracterized eugenics because eugenics is not about 'individual thriving'.[73] As Wilson notes, 'eugenics is primarily a view about overall societal, populational, or "racial" improvement'.[74]

Whether or not PGT is a eugenic practice is the subject of disagreement and this lack of consensus can also be seen in the context of pre-natal diagnosis.[75] Against this background, Tom Shakespeare concludes that pre-natal diagnosis is 'not straightforwardly eugenic' and this can also be applied to PGT.[76] That being said, some academics suggest that 'the word "eugenic" may not be helpful, since it tends to be wheeled out in a discussion-stopping way or simply to have this effect when it

---

70 Widespread use of embryo selection may nevertheless have a 'collectively beneficial' impact on the gene pool: Anomaly J (2018) Defending Eugenics: From Cryptic Choice to Conscious Selection. Monash Bioethics Review 35: 24–35, 34.

71 Anomaly J (2018) Defending Eugenics: From Cryptic Choice to Conscious Selection. Monash Bioethics Review 35: 24–35, 24.

72 Unlike others who use the spectre of eugenics to argue that PGT should not be permitted, Anomaly is not opposed to PGT and in fact argues that 'taxpayers should be willing to finance genetic testing'. Anomaly J (2018) Defending Eugenics: From Cryptic Choice to Conscious Selection. Monash Bioethics Review 35: 24–35, 31.

73 Wilson RA (2019) Eugenics Undefended. Monash Bioethics Review 37: 68–75, 70.

74 Ibid.

75 E.g. Fiona Kumari Campbell refers to pre-natal screening as 'laissez-faire eugenics': Campbell FK (2000) Eugenics in Disguise? Law, Technologies and Negotiating the 'Problem' of 'Disability'. Australian Feminist Law Journal 14: 55–70, 57.

76 Shakespeare T (2013) Disability Rights and Wrongs Revisited. Routledge, 136.

appears or is sensed on the horizon'.[77] Given its history, eugenics is a highly emotive expression[78] that may give rise to unnecessary confusion. Consequently, I suggest that it should be omitted from the language of PGT.

Thus far, I have considered the expressivist objection, the loss of support objection and the issue of eugenics; all of which have been put forward as possible reasons to prohibit the use of PGT to select against disability. Having examined each argument in turn and concluded that none present a sufficient justification for curtailing the use of PGT to prevent disabling conditions, I now turn to the arguments supporting the use of PGT for this purpose.

5.2.4 Procreative Beneficence

Julian Savulescu's principle of procreative beneficence is the idea that prospective parents 'should select the child, of the possible children they could have, who is expected to have the best life, or at least as good a life as the others'.[79] Savulescu draws an analogy between the 'should' in 'should choose the best child' and the 'should' in 'you should stop smoking'.[80] Telling someone that they should stop smoking does not impose an obligation on them to do so and does not condemn them should they choose to continue smoking. It merely suggests that there are sound reasons (namely health) to stop smoking, which would make it a good idea. In a similar vein, suggesting that parents 'should' have the best children implies that there are good reasons to have children with the best life chances; however, parents may choose not to have such children and no countervailing obligation exists.

In a later paper, Savulescu and Guy Kahane offer an improved definition that addresses any confusion caused by the word 'should'.[81] They state that prospective

---

[77] Scott R, 'Choosing between Possible Lives: Legal and Ethics Issues in Preimplantation Genetic Diagnosis' (2006) 26 *Oxford Journal of Legal Studies* 153, 162; Wilkinson S, 'Eugenics, Embryo Selection and the Equal Value Principle' (2006) 1 *Clinical Ethics* 46, 47; Veit W et al. (2021) Can 'Eugenics' be Defended? Monash Bioethics Review 39: 60–67, 61; Wilkinson S (2008) "Eugenics Talk" and the Language of Bioethics. Journal of Medical Ethics 34: 467–471, 471. Gillon also make this point, albeit in the context of abortion; see Gillon R, 'Eugenics, Contraception, Abortion and Ethics' (1998) 24 *Journal of Medical Ethics* 219.

[78] Wilkinson S, Garrard E, 'Eugenics and the Ethics of Selective Reproduction' (Report, Keele University, 2013), 8.

[79] Savulescu J, 'Procreative Beneficence: Why we should select the best children' (2001) 15 *Bioethics* 413, 415. It should be noted that Savulescu's principle applies to both disabilities and abilities; however, this chapter is only concerned with the principle as it applies to selection against disability. The application of Procreative Beneficence to selection for abilities will be considered in the next chapter.

[80] Savulescu J, 'Procreative Beneficence: Why we should select the best children' (2001) 15 *Bioethics* 413–426, 415.

[81] For example, Robert Sparrow argues that Savulescu moves between the claim that prospective parents have good reasons to select the best children, and the much stronger claim that they have an obligation to do this. See Sparrow R, 'Procreative Beneficence, Obligation and Eugenics' (2007)

parents 'have a significant moral reason to select the child, of the possible children they could have, whose life can be expected... to go best or at least not worse than any of the others'.[82] Therefore, the principle 'tells parents to aim to have the child who, given her genetic endowment, can be expected to enjoy the most well-being in her life'.[83] The authors also make clear that procreative beneficence 'in no way suggests that parents should have no child at all rather than one who is less advantaged than other existing children'.[84] Thus in the context of selection against disabling conditions, procreative beneficence suggests that embryos free from disabling conditions should be selected. This is due to the negative impact disabling conditions may have on a person's wellbeing. If, however, all the embryos created will develop a particular disabling condition, procreative beneficence would support selection of the embryo with the least serious disabling condition; that is, the disabling condition likely to cause the fewest limitations to a person's life.

5.2.4.1 Critiquing Procreative Beneficence

Andrew Hotke argues that the principle of procreative beneficence fails because, contrary to Savulescu,[85] morality does not require us to do what we have most reason to do.[86] Savulescu's understanding of morality leads him to state that 'a person who has good reason to have the best child is morally required to have the best child'.[87] Hotke, on the other hand, argues that if Savulescu were correct it would have far-reaching implications and lead to acts which are currently considered acceptable being classed as morally wrong.[88] Hotke illustrates this with reference to a person choosing to skip breakfast. He asserts that assuming there is more reason to eat breakfast than to forgo breakfast, that person would have acted in a manner that is morally wrong. Hotke rightly concludes that this 'is absurd'.[89] Instead, Hotke, taking his lead from Derek Parfit, suggests that 'what we ought practically to do is what we have most reason to do... What we ought morally to do is something

---

3(3) *Genomics, Society and Policy* 43, 44. This point is also made by Rosamund Scott: Scott R (2007) Why Parents have No Duty to Select 'the best' Children. Clinical Ethics 2(3): 149–154, 150.

[82] Savulescu J, Kahane G, 'The Moral Obligation to Create Children with the Best Chance of the Best Life' (2009) 23 *Bioethics* 274, 274.

[83] Ibid 275.

[84] Ibid.

[85] Savulescu J 'Procreative Beneficence: Why We Should Select the Best Children', (2001) 15 *Bioethics* 413, 415.

[86] Hotke A, 'The Principle of Procreative Beneficence: Old Arguments and a New Challenge' (2014) 28 *Bioethics* 255.

[87] Savulescu J 'Procreative Beneficence: Why We Should Select the Best Children', (2001) 15 *Bioethics* 413, 415.

[88] Hotke A 'The Principle of Procreative Beneficence: Old Arguments and a New Challenge' (2014) 28 *Bioethics* 255, 261.

[89] Ibid.

separate'.[90] Thus Hotke observes that Savulescu has 'dubiously conflated' moral reasons with practical reasons.[91] Is this criticism really the crucial blow to procreative beneficence that Hotke suggests it is? While Savulescu's understanding of morality may understandably be considered problematic and Hotke has a valid point, Hotke's criticism does not cut to the core of procreative beneficence. To explain why, it is necessary to look at beneficence.

Beauchamp and Childress state that morality requires us to 'contribute' to the welfare of others.[92] Further, they define the principle of beneficence as 'a statement of moral obligation to act for the benefit of others'.[93] As procreative beneficence is founded on the principle of beneficence, to act in a procreatively beneficent manner and thus act morally, prospective parents simply need to be acting for the benefit of their future child. A prospective parent may have practical reasons for selecting a child without a disabling condition, for example, a child who will not require regular hospital appointments would be less disruptive to their parents' lifestyle. Similarly, a child who does not require expensive medical treatment would be less of a financial strain on their parents. However, these practical reasons are not for the benefit of the child and are therefore not moral reasons. To select the embryo that will become a child without an intellectual disability because they may struggle less at school and will have a selection of career paths available to them would be acting for the benefit of the child and thus a moral reason. Therefore, while Hotke may have a valid criticism of Savulescu's understanding of morality, this does not necessarily affect the validity of procreative beneficence.

Savulescu argues that procreative beneficence reflects one of the basic principles of parenting, 'that parents should give their children the best possible start in life'.[94] He suggests that giving a child the best start in life may be understood to include choosing a child who begins life with the best chance of leading the best life. However, the idea of the 'best life' has left procreative beneficence open to criticism. For example, Michael Parker asserts that the principle of procreative beneficence is 'underdetermining' and therefore void of real world application.[95] He suggests that this is because what counts as a good life, or the best life, is made up of 'rich and complex concepts' and it would be impossible to coherently rank possible lives in the way required by procreative beneficence.[96] Savulescu's response is that Parker's objection relates not specifically to procreative beneficence, but to placing values on

---

[90] Ibid. See also Parfit D, *On What Matters* (Oxford University Press, 2011).

[91] Hotke A (2014) The Principle of Procreative Beneficence: Old Arguments and a New Challenge. 28 *Bioethics* 255, 261.

[92] Beauchamp TL, Childress JF, *Principles of Biomedical Ethics* (Oxford University Press, 6th ed, 2009) 197.

[93] Ibid.

[94] Savulescu J, 'Procreative Beneficence: Reasons to not have Disabled Children' in Skene L, Thompson J (eds), *The Sorting Society: The Ethics of Genetic Screening and Therapy* (Cambridge University Press, 2008) 51, 63.

[95] Parker M, 'The Best Possible Child' (2007) 33 *Journal of Medical Ethics* 279, 281.

[96] Ibid.

lives.[97] Savulescu notes that evaluations about lives are constantly being made, for example, hospitals exist 'because disease and injury are bad, not because they are good or because we are indifferent to them'.[98] Therefore, in the context of disabling conditions, it may be possible, at least tentatively, to say that embryos which will become individuals with serious impairments (for example, those that cause chronic pain) may be ranked as less likely to lead the best possible life than embryos which will develop into individuals with minor impairments.

In their restatement of the principle of procreative beneficence, Savulescu and Kahane refer to parents having a 'significant moral reason' for selecting the best child. This reason appears to be fairly straightforward; that selecting the child with the best life chances leads to the better outcome.[99] The authors point out that this 'better outcome' may be 'person affecting', in that a more advantaged child will gain a greater benefit by being brought into existence; or 'non-person affecting', meaning that even if the outcome is not better for the specific child created, selecting the child with the best prospects would make for a better overall outcome.[100] The best outcome is the outcome that maximizes wellbeing and as the authors note, 'when we make decisions the option we should choose is the one which maximizes expected value'.[101] However, it is unclear what expected value means in the context of having children. Further, how is this 'value' to be judged? Savulescu and Kahane note that the idea of pursuing maximum wellbeing in future children is the same as striving for the best possible outcomes in our own lives.[102] Therefore, in the context of selection against disabling conditions, procreative beneficence supports maximizing wellbeing by selecting against disabling conditions that reduce wellbeing, for example, by causing pain and suffering.

The principle of procreative beneficence does not distinguish between serious disabling conditions and less serious disabling conditions. This is because disabling conditions are only relevant insofar as they have an impact on wellbeing.[103] Under the interactional model, if a genetic condition is described as a disabling condition, the interaction between the impairment and the environment into which the child will be born will have a negative impact on that child's wellbeing.[104] Therefore, procreative beneficence would support the use of PGT to select against disabling conditions, even if trivial, because such conditions reduce wellbeing and the desired outcome is to maximize wellbeing.

---

[97] Julian Savulescu, 'In Defence of Procreative Beneficence' (2007) 33 *Journal of Medical Ethics* 284, 284.

[98] Ibid 285.

[99] Savulescu J, Kahane G (2009) The Moral Obligation to Create Children with the Best Chance of the Best Life. 23 *Bioethics* 274, 277.

[100] Ibid.

[101] Ibid 278.

[102] Savulescu J, Kahane G (2009) The Moral Obligation to Create Children with the Best Chance of the Best Life. *Bioethics* 23: 274, 279.

[103] Savulescu and Kahane make this point in relation to disease, stating that 'disease itself is morally significant only in so far as it reduces wellbeing' (ibid 277).

[104] A similar conclusion can be reached if one uses Savulescu's welfarist model of disability.

Procreative beneficence is an essentially simple concept with its roots firmly in utilitarian theory. People should do what they have good reason to do and selecting an embryo that will develop into a child without a disabling condition, over an embryo that will develop into a child with a disabling condition, makes sense if it maximizes that child's[105] wellbeing.[106] Accordingly, if prospective parents wish to act in a procreatively beneficent manner, the regulatory framework should permit them to do so. To be clear, this should not be seen as imposing an obligation on prospective parents to make particular choices, but rather one way in which they may exercise their reproductive liberty. Having explored the principle of procreative beneficence, this chapter moves on to a second argument that may support the use of PGT to select against disabling conditions.

5.2.5 The Open Future Argument

This section considers Joel Feinberg's paper 'The Child's Right to an Open Future'.[107] It asks whether giving a child an 'open future' is an appropriate reason to allow parents to use PGT to select against disabling conditions. In his paper, Feinberg divides rights into three categories. The first category contains rights that are common to both adults and children,[108] such as the right not to be subjected to inhuman treatment. The second category comprises rights that only belong to adults, for example, the right to vote.[109] The final category of rights are those rights specific to children and it is this category that is most relevant to the present discussion. Feinberg separates this category into two sub-categories: 'dependency rights' and 'rights in trust'.[110] Dependency rights stem from a child's dependence on others and includes the right to be provided with food and shelter.[111] 'Rights in trust' are similar to adult rights insofar as their exercise involves the making of informed decisions. However, as the child is incapable of exercising them, they 'are to be *saved* for the child until he is an adult'.[112] Feinberg argues that it is possible for adults to violate these 'rights in trust' so

[105] Cf Parfit who argues that it is not a particular child's wellbeing which is maximised because a different child is chosen. See Parfit D, *Reasons and Persons* (Oxford University Press, 1986) 359. This argument is explored further in Chap. 7.

[106] Savulescu and Kahane note that if there are good reasons to select an embryo that will become a child with a disability, it may be morally permissible to do so. This argument will be considered further in Chap. 7.

[107] Feinberg J (1980) The Child's Right to an Open Future. In Aiken W, LaFollette H (eds), *Whose Child? Children's Rights, Parental Authority and State Power* (Littlefield, Adams & Co) 124.

[108] Ibid 125.

[109] Ibid.

[110] Ibid.

[111] Ibid.

[112] Ibid.

> that when the child is an autonomous adult, certain key options will already be closed to him. His right while he is still a child is to have these future options kept open until he is a fully formed self-determining adult capable of deciding among them.[113]

Feinberg refers to these 'rights in trust' collectively as the 'right to an open future'.[114]

The right to an open future may be considered as a collection of 'anticipatory autonomy rights'.[115] Feinberg suggests that the word autonomy may be understood in two ways; first, as referring to the 'capacity' of a person to exercise control over their own actions, and second, as self-determination.[116] Therefore, in situations where the state has to intervene because the rights of a parent have come into conflict with those of their child, the state may argue that the child's right to an open future will be violated if 'certain crucial and irrevocable decisions' that would determine the course of the child's life are made before that child has attained 'the *capacity of self-determination* himself'.[117]

Feinberg's paper focuses on the religious rights of parents when they come into conflict with the rights in trust of their children. He illustrates this conflict by reference to *Wisconsin v Yoder*,[118] where the United States Supreme Court permitted Amish families to withdraw their children from school two years before finishing their compulsory education. By denying their children the chance to finish high school, Amish parents were limiting the opportunities that would be available to their children in the future. Further, if one accepts that the State of Wisconsin had sound reasons for requiring all children to receive a complete high school education, it is unclear why the Amish children were denied this minimum entitlement.[119] As Dena Davis notes,

> none of the judges squarely faced the question of whether the liberal democratic state owes all its citizens, especially children, a right to a basic education that can serve as a building block if the child decides later in life that she wishes to become an astronaut, a playwright, or perhaps [a lawyer].[120]

Furthermore, she points out that by denying their children a full high school education, Amish parents 'are virtually ensuring that their children will remain housewives and agricultural labourers'.[121] Feinberg himself similarly states that an 'education that renders a child fit for only one way of life forecloses irrevocably his other

---

[113] Ibid 126.

[114] Ibid.

[115] Ibid.

[116] Ibid 142–143.

[117] Ibid 143 (emphasis in original).

[118] 406 US 205 (1972).

[119] Davis DS (2010) Genetic Dilemmas: Reproductive Technology, Parental Choices, and Children's Futures. Oxford University Press, New York 29.

[120] Davis DS, 'Genetic Dilemmas and the Child's Right to an Open Future' (1997) 27(2) *Hastings Center Report* 7, 10.

[121] Ibid.

options'.[122] In this way, the Amish children in *Yoder* were effectively being denied their right to self-determination.

Thus far, my discussion of the open future argument has centred on the rights of existing children. This begs the question; can the open future argument be applied to embryos? There appears no reason the open future argument cannot be applied to an embryo. When prospective parents create embryos in vitro with a view to them becoming future children, it is possible to say that those embryos have the right to an open future. However, unlike children who have already been born, an embryo can lose its right to an open future. This is because it is the parental intention to implant the embryo and then carry it to term which gives it such a right. This right may be understood as a 'conditional right'[123]; conditional upon the intention that the embryo will exist as a person at some point in the future. Therefore, if a prospective parent chooses to discard an embryo, that embryo's right to an open future ceases because there is no longer an intention for it to become a child.

The application of the open future argument in the context of PGT-M has been illustrated by Janet Malek and Judith Daar in the context of autosomal recessive polycystic kidney disease (ARPKD).[124] ARPKD causes cysts to form on the kidneys and liver. While the symptoms of the disease may vary in their severity, in the most serious cases a child will die within the first year of their life.[125] Children who survive beyond this will experience a range of symptoms including enlarged kidneys and compromised kidney function, and will require a kidney or liver transplant by the time they reach adolescence or early adulthood.[126] Malek and Daar argue that ARPKD is a barrier to a child's open future because it may affect the choices available to the child. Even though the child's intellectual capacities are not affected by the condition and thus the child can choose between the options available to them, 'the capacity to make such choices is less meaningful if the range of options actually available is severely limited'.[127] Consequently, they suggest that there may be a duty on prospective parents to use PGT to select against disabling conditions in circumstances where ART is being used and they are aware of the risk of passing on a 'serious genetic' condition to any child born.[128]

The problem with imposing a duty on prospective parents to widen their future child's open future is that there is no logical reason for limiting this duty to PGT-M. Accordingly, such a duty could be said to create an obligation to test embryos

---

[122] Feinberg J, 'The Child's Right to an Open Future' in William Aiken and Hugh LaFollette (eds), *Whose Child? Children's Rights, Parental Authority and State Power* (Littlefield, Adams & Co, 1980) 124, 132.

[123] Wilkinson S (2005) 'Designer Babies', Instrumentalisation and the Child's Right to an Open Future. In: Athanassoulis N (ed) Philosophical Reflections on Medical Ethics. Palgrave Macmillan, London, 44–69, 60.

[124] Malek J, Daar J (2012) The Case for a Parental Duty to Use Preimplantation Genetic Diagnosis for Medical Benefit. American Journal of Bioethics, 12(4): 3–11.

[125] Ibid 3.

[126] Ibid 3–4.

[127] Ibid 6.

[128] Ibid 10.

for any advantageous traits that have the potential to expand the future child's open future.[129] The imposition of such a duty is also problematic because it intrudes on the reproductive liberty of prospective parents. As discussed in the previous chapter, interference with reproductive liberty is only permissible for the avoidance of harm. Therefore, unless it is possible to show that actual harm (as opposed to speculative harm)[130] will result if a duty to select the embryo with the most open future is not imposed, restricting the reproductive liberty of prospective parents is impermissible. Leaving aside the broader problems associated with imposing a duty to use PGT on prospective parents, Malek and Daar's formulation also gives rise to the intractable question of how to define what qualifies as a 'serious' genetic condition[131]; this will be discussed later in this chapter. While it is problematic to impose on prospective parents an obligation to use PGT to select against disabling conditions, it is possible to say that giving a future child an open future is an acceptable exercise of prospective parents' reproductive liberty.

5.2.5.1 Critiquing the Open Future Argument

Feinberg's open future argument has not gone uncriticised. Claudia Mills asserts that 'it is both impossible and undesirable to try to provide children with an "open" future in any meaningful sense'.[132] Further, she argues that 'it is difficult to decipher what it even means to have "open" or "closed" options, and to figure out how options are to be counted and differentiated'.[133] Using *Yoder* as an example, Mills notes that an Amish child may leave the Amish community when they reach adulthood and would then be free to pursue careers that were previously closed to them. While it may be more difficult for such an adult to pursue, for example, a career in medicine, Mills notes that it would by no means be impossible.[134] However, the same cannot be said about a child with a severe disabling condition and thus Mills' critique does not apply in the context of selection against disability. Consider an embryo that becomes a child

[129] De Melo-Martin I (2012) A Parental Duty to Use PGD: More Than We Bargained For? American Journal of Bioethics 12(4): 14–15, 15.

[130] John Harris notes that if speculative or future harms were sufficient, imaginative dictators could invent all manner of anticipated harms in order to erode liberty. Harris J (2007) Enhancing Evolution: The Ethical Case for Making Better People. Princeton University Press, New Jersey, 74.

[131] See also Sayres LC and Magnus D (2012) Duty-Free: The Non-Obligatory Nature of Preimplantation Genetic Diagnosis. American Journal of Bioethics, 12(4): 1–2, 1.

[132] Mills C, 'The Child's Right to an Open Future?' (2003) 34 *Journal of Social Philosophy* 499, 499.

[133] Ibid. A similar point is made by Stephen Wilkinson who notes that 'it is hard to say what counts as a more or less open future because it is hard to come up with a non-arbitrary way of individuating and counting options'—Wilkinson S (2010) Choosing Tomorrow's Children: the ethics of selective reproduction. Oxford University Press, Oxford, 46.

[134] Ibid 501. Cf Feinberg, who refers to such a change of career as 'difficult to the point of practical impossibility' (Feinberg J, 'The Child's Right to an Open Future' in Aiken W, LaFollette H (eds), *Whose Child? Children's Rights, Parental Authority and State Power* (Littlefield, Adams & Co, 1980) 124, 132).

with a severe intellectual disability. Unlike the Amish child described by Mills, on reaching adulthood, a child with a severe disabling condition does not acquire the ability to pursue options previously closed to them.

In a similar vein to Mills, Robert Sparrow notes that evaluating the value of various sets of options 'requires more than simply counting the number of options – it requires evaluating the *worth* of the options'.[135] Joseph Raz illustrates this point well, stating that '[a] choice between hundreds of identical and identically situated houses is no choice, compared with a choice between a town flat and a suburban house'.[136] However, Feinberg's argument does not suggest that the value of choices is not important, it simply requires that a child have enough options available to them, so that on reaching adulthood they are in a position to evaluate the various choices for themselves. A child with a severe intellectual disability not only has a limited number of options available, they may also lack the ability to evaluate those options and arrive at a meaningful decision on their value.[137] Therefore, in the context of selection against disabling conditions, Sparrow's critique of Feinberg does not appear to be relevant. By permitting the use of PGT for the purposes of selection against disabling conditions, prospective parents are given the freedom to provide their child with an open future, not just in terms of the number of choices available to their child, but also in respect of the richness of those choices.

Feinberg's open future argument provides a good basis for permitting prospective parents to use PGT to select against disabling conditions. As some commentators have noted, providing a child with a completely open future is an impossibility, as certain choices or traits foreclose certain options.[138] However, even Feinberg accepts that what is important is 'the form of the particular rights in question and not their specific content'.[139] It is thus possible to say that future children have a right to an open future, in the sense that they should have a range of future options available to them and the capacity to decide between these options. In Chap. 4, I said that prospective parents' reproductive liberty extends to gaining information, through the use of PGT, about the embryos that they have created. By selecting against disabling conditions, prospective parents can provide their future child with more options than they may otherwise have had and thus give them as open a future as possible.[140]

---

[135] Sparrow R, 'A Child's Right to a Decent Future? Regulating Human Genetic Enhancement in Multicultural Societies' (2012) 4 *Asian Bioethics Review* 355, 365.

[136] Raz J, *The Morality of Freedom* (Oxford University Press, 1988) 375.

[137] The ability to weigh up a set of options may be considered a 'general-purpose means'—a trait that would prove valuable to a person no matter that kind of life they choose to lead. This idea is explored in the next chapter.

[138] Sparrow R, 'A Child's Right to a Decent Future? Regulating Human Genetic Enhancement in Multicultural Societies' (2012) 4 *Asian Bioethics Review* 355, 365.

[139] Feinberg J (1980) The Child's Right to an Open Future. In: Aiken W, LaFollette H (eds), *Whose Child? Children's Rights, Parental Authority and State Power* (Littlefield, Adams & Co) 124, 126.

[140] An alternative perspective on the term 'open future' is put forward by Rosemarie Garland-Thomson, who suggests that 'disability contributes a narrative of a genuinely open future, one not controlled by the objectives, expectations, and understandings of the present'. Rather than having a metaphorical road map for the future, a child with a disabling condition who, for example, is

Therefore, if prospective parents want to protect their future child's right to an open future, the regulatory framework for embryo testing should support this.

5.2.6 *Late Onset and Susceptibility Conditions*

This chapter has thus far considered the issue of selection against disabling conditions in the context of disabling conditions that will affect the future child from a very early age. This section considers whether the use of PGT should be permitted for late onset and susceptibility conditions. Late onset conditions present interesting issues because a person will have a perfectly healthy life until the condition manifests itself. Susceptibility conditions, which are also known as lower penetrance conditions, present an additional issue as there is no guarantee that a person will ever contract the condition, just an above average chance that this will be the case.

Consider a late onset condition such as Huntington's Disease, which has an average age of onset of 40 years old.[141] Therefore, an individual would be able to have a healthy childhood, a fulfilling career and even start a family[142] prior to experiencing the onset of symptoms. Unlike susceptibility conditions, late onset conditions like Huntington's Disease *will* affect the individual at some point during their life.[143] Should age of onset be a relevant consideration in deciding whether to permit PGT for late onset conditions? An empirical study by Felicity Boardman and Corinna Clark indicated that people who are diagnosed with late onset conditions perceive their quality of life as lower compared to those diagnosed with early onset conditions.[144] Boardman and Clark reported that over three quarters of individuals diagnosed with adult-onset spinal muscular atrophy (type 4) did not consider themselves to have a good level of wellbeing. On the other hand, 76% of individuals with spinal muscular atrophy (type 2), which has an age of onset of between six to eighteen months, considered themselves to have a high level of wellbeing.[145] This

---

expected to die young, may be free to see where life takes them. As Garland-Thomson puts it, they are free 'from the burden of using the present as an investment in a future'. See Garland-Thomson R (2012) 'The Case for Conserving Disability'. Journal of Bioethical Inquiry 9: 339–355.

[141] Myers RH, 'Huntington's Disease Genetics' (2004) 1 *NeuroRx* 255, 255.

[142] Late onset conditions carry a further complication; that an affected parent will die prior to the child reaching adulthood, or become unable to care for the child due to their condition. See generally, Spriggs M, 'Genetically Selected Baby Free of Inherited Predisposition to Early-onset Alzheimer's Disease' (2002) 28 *Journal of Medical Ethics* 290.

[143] Some conditions which manifest themselves later in life are also susceptibility conditions, for example, genes that predispose a person to certain types of cancer in later life. I will explore these within the discussion of susceptibility conditions.

[144] Boardman FK, Clark CC (2022) What is a 'Serious' Genetic Condition? The Perceptions of People Living with Genetic Conditions. European Journal of Human Genetics 30: 160–169. https://doi.org/10.1038/s41431-021-00962-2.

[145] Ibid 163.

data should prompt us to question our assumptions about late onset conditions[146] and suggests that an emphasis on age of onset needs to give way to considerations of what life will be like once the condition manifests itself. As the experiences of people living with disabling conditions are important to the way in which society understands such conditions, this study highlights that late onset conditions have a significant negative impact on wellbeing.

Bonnie Steinbock has suggested that age of onset should be a consideration, because a cure might be developed in the decades prior to the onset of symptoms.[147] Conversely, Laura Purdy asserts that unless a cure is imminent, 'feebly hoping for a cure is both morally questionable and strangely relaxed about the welfare and happiness of one's offspring'.[148] Similarly, John Robertson argues that the 'fact that the disease state does not occur until much later in life ... should not be morally significant'.[149] This is because even if a cure is developed during the future child's lifetime, the use of PGT has not harmed the child or society at large. Accordingly, the argument that a cure may be developed would not constitute a sound moral basis for prohibiting PGT for late onset conditions. As the freedom to test for a late onset condition may affect the prospective parents' decision as to whether to reproduce, it falls within the scope of reproductive liberty. Consequently, prohibiting testing for late onset conditions is not a justifiable restriction on the reproductive liberty of prospective parents. Further, as Isabel Karpin and Kristin Savell note, 'as the age of forty is an age at which many people still have young, dependent children and are seen as being at the beginning of midlife, to develop a life-threatening or fatal illness at this time is an extremely serious matter'.[150] Taking into account the likelihood of debilitating symptoms and the potential impact of living with the knowledge that a disabling condition will manifest itself during mid-life, there seems to be consensus that parents should have the freedom to test for late onset conditions.

I now turn to lower penetrance, or susceptibility conditions. Should parents have the freedom to use PGT to test for susceptibility conditions, even though their future child may never suffer from the condition? When the United Kingdom's Human Fertilisation and Embryology Authority (HFEA) licensed PGT for inherited cancer susceptibility, namely susceptibility to breast, ovarian and bowel cancers. During the consultation process, the HFEA that susceptibility to cancer was different from previously licensed conditions, due to a combination of three factors. First, the lower

---

[146] Newson AJ, Dive L (2022) Taking Seriousness Seriously in Genomic Health. European Journal of Human Genetics 30:140–141. https://doi.org/10.1038/s41431-021-01002-9.

[147] Steinbock B, 'Prenatal Testing for Adult-onset Conditions: cui bono?' (2007) 15 (Supplement 2) *Reproductive Biomedicine Online* 38, 40.

[148] Purdy L, 'Too Late to Matter? Preventing the Birth of Infants at Risk for Adult-Onset Disease or Disability' in D Christopher Ralston and Justin Ho (eds), *Philosophical Reflections on Disability* (Springer, 2010) 109, 115.

[149] Robertson JA (2003), 'Extending Preimplantation Genetic Diagnosis: the ethical debate - Ethical issues in new uses of preimplantation genetic diagnosis' 18 *Human Reproduction* 465, 467.

[150] Karpin I, Savell K (2012), *Perfecting Pregnancy: Law, Disability, and the Future of Reproduction* (Cambridge University Press) 292.

penetrance of inherited cancers, second, the late onset of the condition and lastly, the existence of prophylactic surgery and effective treatment.[151]

As it has already been accepted that age of onset should not matter, the question remains, should PGT be permitted for conditions that may never manifest themselves and even if they do, are treatable? Jeanne Snelling notes that the penetrance of a genetic condition should not be the sole factor in deciding whether to permit PGT for susceptibility conditions as the phenotypic traits associated with the condition are of equal importance.[152] Further, Snelling states that it would be unfair to accuse parents seeking to select against susceptibility conditions of 'seeking "perfection". They are simply trying to avoid their child's future being clouded by the threat of serious genetic disease'.[153] One justification for permitting selection against susceptibility conditions is that it permits parents to act in a procreatively beneficent manner. As previously discussed, the principle of procreative beneficence does not distinguish between serious and less serious disabling conditions and would support the use of PGT for trivial conditions. As cancer, for example, is by no means a trivial condition, permitting PGT for the avoidance of inherited cancer predisposition would allow parents to act in a manner that is procreatively beneficent. Even though a susceptibility condition may not actually manifest itself, if it does, it will result in a reduction in an individual's wellbeing. Prospective parents should therefore be permitted to test embryos for susceptibility conditions and to select the embryos that are free of genetic conditions that will reduce the future child's wellbeing.

5.3 The Australian Regulatory Framework

As discussed in chap. 2, the National Health and Medical Research Council's *Ethical Guidelines on the Use of Assisted Reproductive Technology in Clinical Practice and Research 2017* only permit the use of PGT to test for and select against genetic conditions 'that would severely limit the quality of life' of the future child.[154] However, the *Assisted Reproductive Treatment Act 1988* (SA) s 9(1)(c)(iii) restricts the use of PGT to testing for *serious* genetic defects, *serious* diseases and *serious* illnesses, while the *Human Reproductive Technology Act 1991* (WA) s 14(2b)(a)(ii) limits the use of PGT to cases where 'there is a significant risk of a serious genetic abnormality or disease'. Interestingly, neither piece of legislation provides a definition of what qualifies as a

---

151 Human Fertilisation and Embryology Authority, 'Choices and Boundaries: Should People be Free to Select Embryos Free from an Inherited Susceptibility to Cancer?' (10 November 2005).

152 Snelling J (2008), 'Preimplantation Genetic Diagnosis for Susceptibility Conditions: a new frontier or a logical extension?' 16 *Journal of Law and Medicine* 263, 266.

153 Ibid 273.

154 National Health and Medical Research Council, *Ethical Guidelines on the Use of Assisted Reproductive Technology in Clinical Practice and Research 2017* ('*NHMRC Guidelines*') [8.15.1].

'serious' condition, despite calls for guidance on this concept.[155] Further, when the term 'serious' was previously included in the NHMRC Guidelines, the Guidelines acknowledged that 'what counts as a serious genetic condition is controversial'.[156]

While Australian legislators have been reluctant to provide guidance on the meaning of the word 'serious' in the context of disabling conditions, the UK's HFEA has listed a number of factors that should be taken into account by clinics in deciding whether to authorise PGT-M or PGT-SR for 'a gene, chromosome or mitochondrion abnormality'.[157] In their Code of Practice, the HFEA provides that clinics should consider the following factors in deciding whether such an abnormality will result in the future child developing 'a serious physical or mental disability, a serious illness' or 'a serious medical condition'[158]

(a) the views of the people seeking treatment in relation to the condition to be avoided, including their previous reproductive experience
(b) the likely degree of suffering associated with the condition
(c) the availability of effective therapy, now and in the future
(d) the speed of degeneration in progressive disorders
(e) the extent of any intellectual impairment
(f) the social support available, and
(g) the family circumstances of the people seeking treatment.[159]

Similar considerations are found in the NHMRC Guidelines for the assessment of whether a genetic condition will constitute a severe limitation to the future child's quality of life.[160] The inclusion of factors such as the availability of social support and the family circumstances of the prospective parents, suggest that the seriousness of a disabling condition depends on more than just the symptomatic elements of the condition and is tied up with social factors. This was also highlighted by Boardman

[155] Savell K, Karpin I (2008), 'The Meaning of "serious disability" in the Legal Regulation of Prenatal and Neonatal Decision-making' 16(2) *Journal of Law and Medicine* 233, 245. The Western Australian government has provided guidance on the factors that will be considered by the Reproductive Technology Council's PGT Committee when determining of seriousness. See, Reproductive Technology Unit: Department of Health, Government of Western Australia (2025) Preimplantation Genetic Testing Applications to the Reproductive Technology Council: Guidelines for Clinics. https://www.rtc.org.au/wp-content/uploads/2025/03/20250328-Guidelines-for-clinics-REDCap-PGT-applications.pdf Accessed 24 July 2025.

[156] National Health and Medical Research Council, *Ethical Guidelines on the Use of Assisted Reproductive Technology in Clinical Practice and Research 2007* 55 [12.1].

[157] *Human Fertilisation and Embryology Act 1990* (UK) s 1ZA(1)(b).

[158] *Human Fertilisation and Embryology Act 1990* (UK) s 1ZA(2).

[159] Human Fertilisation and Embryology Authority (2021) Code of Practice (9th edition) 10.9. These considerations also apply where PGT-A is used at the same time as PGT-M or PGT-SR (10.8).

[160] National Health and Medical Research Council, *Ethical Guidelines on the Use of Assisted Reproductive Technology in Clinical Practice and Research 2017* [8.16.1].

and Clark's empirical study of individuals with 'clinically serious' disabling conditions, which found that the environmental restrictions which participants experienced played a significant role in their negative lived experiences.[161]

The meaning of the word serious has also been debated in the context of abortion in the UK.[162] In 2003, British curate Joanna Jepson applied for permission to proceed with a claim for judicial review of a decision to allow a late term abortion for a bilateral cleft palate.[163] Jepson argued that a cleft lip and palate did not qualify as a serious handicap for the purposes of s 1(1)(d) of the *Abortion Act 1967* (UK), and therefore the doctors who terminated the pregnancy committed a crime.[164] Jepson initially wrote to West Mercia Police asking them to prosecute the doctors who authorised the procedure, but after investigating the matter the police concluded that no crime had been committed.[165] After Jepson was granted permission to proceed with her claim for judicial review, the police agreed to reopen the investigation, but once again found the doctors' actions were justified.[166] Even though Jepson did not go ahead with the action for judicial review, the case provides an example of the difficulties caused by the word serious in the context of reproductive choices.[167]

In 2006, the Human Genetics Commission in the UK noted that while a possible method of defining serious conditions would be by reference to a list of conditions considered serious, this would not be an adequate way forward as 'this approach fails to recognise that quality of life judgments are subjective'.[168] This subjectivity is encapsulated by the Human Genetics Commission's acknowledgement that 'judgements about seriousness depend heavily upon context, including the parents' existing

[161] Boardman FK, Clark CC (2022) What is a 'Serious' Genetic Condition? The Perceptions of People Living with Genetic Conditions. European Journal of Human Genetics 30: 160–169, 167. https://doi.org/10.1038/s41431-021-00962-2.

[162] Despite the prevalence of the term serious in the context of reproductive choice, Kleiderman et al. found that the 'most clearly articulated definition' of the word serious in the healthcare context is provided by the United States Food and Drug Administration. The definition applies to the use of investigational new drugs and provides that 'serious disease or condition means a disease or condition associated with morbidity that has substantial impact on day-to-day functioning. Short-lived and self-limiting morbidity will usually not be sufficient, but the morbidity need not be irreversible, provided it is persistent or recurrent. Whether a disease or condition is serious is a matter of clinical judgment, based on its impact on such factors as survival, day-to-day functioning, or the likelihood that the disease, if left untreated, will progress from a less severe condition to a more serious one'. Kleiderman E, Ravitsky V, Knoppers BM (2020) The 'Serious' Factor in Germline Modification. Journal of Medical Ethics 45: 508–513.

[163] *Jepson v The Chief Constable of West Mercia Police Constabulary* [2003] EWHC 3318 (Admin).

[164] Ibid [3].

[165] Ibid [4].

[166] Dyer C (2005) Doctors who performed late abortion will not be prosecuted. BMJ 30: 688.

[167] The very inclusion of the term 'seriously handicapped' has also been the subject of an application for judicial review—*R(Crowter) v Secretary of State for Health and Social Care* [2021] EWHC 2536.

[168] Human Genetics Commission, 'Making Babies: Reproductive Decisions and Genetic Technologies' (January 2006) 36–37 [3.27].

resources and capacities and society's willingness to support those with genetic disorders and their families'.[169] Similarly, Karpin and Savell describe the question of seriousness as more than 'a purely clinical determination, as parents will have views about whether or not the disability being described to them is one with which they feel they can cope'.[170]

The definitional dilemmas presented by the word serious are often illustrated by reference to Down syndrome.[171] For example, whether Down syndrome is a serious genetic condition is a question which makes many people anxious.[172] Timothy Krahn argues that categorising Down syndrome as a serious condition warranting PGT 'risks propounding a misconception of what life is really like for persons with Down's syndrome and their families'.[173] Similarly, Amber Knight and Joshua Miller cite empirical evidence indicating that 99% of people with Down syndrome are 'happy with their lives'.[174] Thus Down syndrome has the potential 'to destabilize a threshold' for serious disabling conditions that is based on the idea that such a condition has a negative impact on the life of the person with the condition.[175]

Considering the subjectivity inherent in the term 'serious', is it necessary for seriousness to be one of the criteria on which access to PGT is based? Karpin and Savell argue that the word 'serious' forms part of the regulatory framework in order to convey 'the idea of a limit, rather than its reality'.[176] However, as I have noted elsewhere, 'the difficulty of drawing a line between the mild end of serious disability and the severe tip of trivial disability, means that the line between the two is at best imprecise and at worst arbitrary and futile'.[177] We must therefore 'be wary

---

[169] Ibid.

[170] Karpin I, Savell K (2012) Perfecting Pregnancy: Law, Disability, and the Future of Reproduction (Cambridge University Press) 275. This point is also made by Michelle Bayefsky and Bruce Jennings –Bayefsky M, Jennings B (2015) Regulating Preimplantation Genetic Diagnosis in the United States: The Limits of Unlimited Selection. Palgrave Macmillan, New York 44.

[171] See generally, Karpin I, Savell K (2012) Perfecting Pregnancy: Law, Disability, and the Future of Reproduction (Cambridge University Press) 277–281; Krahn TM (2011) Regulating preimplantation genetic diagnosis: the case of Down's Syndrome. Medical Law Review: 19(2): 157–191; Knight A, Miller J (2015) Prenatal Genetic Screening, Epistemic Justice, and Reproductive Autonomy. Hypatia 36: 1–21.

[172] Karpin I, Savell K (2012) Perfecting Pregnancy: Law, Disability, and the Future of Reproduction (Cambridge University Press) 277.

[173] Krahn TM (2011) Regulating preimplantation genetic diagnosis: the case of Down's Syndrome. Medical Law Review: 19(2): 157–191, 189.

[174] Knight A, Miller J (2015) Prenatal Genetic Screening, Epistemic Justice, and Reproductive Autonomy. Hypatia 36: 1–21, 11.

[175] Karpin I, Savell K (2012) Perfecting Pregnancy: Law, Disability, and the Future of Reproduction (Cambridge University Press) 278.

[176] Karpin I, Savell K (2012) Perfecting Pregnancy: Law, Disability, and the Future of Reproduction (Cambridge University Press) 277.

[177] De Souza M (2015) The Regulation of Preimplantation Genetic Diagnosis in Australia: Disability and Parental Choice. Journal of Law and Medicine Journal of Law and Medicine 22: 915–933, 932.

of making ethically arbitrary distinctions'.[178] Karpin and Savell suggest that as 'it is not possible to reach objective determinations of "serious disability," that term should instead carry the meaning that it is given by the [prospective parents]'.[179] This approach is sensible because the responsibility for raising a child with a disabling condition falls mainly on the prospective parents.[180] As Michelle Bayefsky and Bruce Jennings observe, the difficulties inherent in defining seriousness may in fact form part of its appeal, as its vagueness 'allows for flexibility in determining when [PGT] is appropriate'.[181] Although a subjective approach to disability is desirable and well supported within the interactional model of disability, the utility of the term 'serious' is still questionable. If the word serious has no settled meaning and conditions such as Down syndrome, that many argue do not meet the seriousness threshold, are considered acceptable for PGT, the use of the term serious appears to be ineffectual. I thus recommended that references to seriousness in the context of selection against disabling conditions be removed from the Australian regulatory framework.

5.4 Conclusion

This chapter has argued that parents should have the freedom to use PGT to test for and select against disabling conditions. The chapter begun by considering the objections to the use of PGT to select against such conditions. I argued that these objections are flawed in a number of ways and thus do not present satisfactory reasons to prevent or even discourage parents from using PGT to select against disabling conditions. As these objections apply equally to all disabling conditions, they do not provide a justification for limiting the use of PGT to serious conditions, as is currently the case in South Australia and Western Australia. Consider once again the expressivist argument. The argument is that selection against disability should not be permitted because it reflects negative attitudes about people with disabling conditions. These attitudes can be expressed both by the prospective parents choosing to undergo IVF with PGT and by the state in permitting PGT. The argument does not discriminate between different degrees of disability, as a negative attitude can just as easily be expressed about a person with a severe disabling condition as it can about a person with a less serious disabling condition. The expressivist argument has been rejected for firstly, failing to distinguish between disabling conditions and persons

[178] Bayefsky M, Jennings B (2015) Regulating Preimplantation Genetic Diagnosis in the United States: The Limits of Unlimited Selection. Palgrave Macmillan, New York 44. Similar concerns have been echoed by Kleiderman et al. who note that the 'undefined nature of "serious" renders its application vague and decision making surrounding [it] … potentially arbitrary.'—Kleiderman E, Ravitsky V, Knopper BM (2020) The 'Serious' Factor in Germline Modification. Journal of Medical Ethics 45: 508–513, 511.

[179] Ibid 347.

[180] Murray T (1996) The Worth of a Child. University of California Press, 138.

[181] Bayefsky M, Jennings B (2015) Regulating Preimplantation Genetic Diagnosis in the United States: The Limits of Unlimited Selection. Palgrave Macmillan, New York 44.

with disabling conditions and secondly, because sending hurtful messages to people with disabling conditions is not an adequate justification for restricting reproductive liberty. In light of this, I propose that the requirement that selection can only be undertaken against serious disabilities should also be rejected.

This chapter has also considered two arguments that support the use of PGT to select against disabling conditions. Like the arguments advanced in opposition to PGT, the arguments in favour of this practice do not distinguish between serious and less serious conditions. This supports the proposition that parents should be permitted to select against any condition identified as disabling. One of the principles considered is that of procreative beneficence; the idea that parents have good reasons to select the child who can be expected to have the best life. In the context of procreative beneficence it was noted that disabling conditions, even if mild, have the potential to reduce wellbeing and therefore selection against such conditions maximizes wellbeing. As it is sensible to select embryos which will begin life with the most wellbeing, prospective parents should be allowed to test for and select against any disabling conditions. It has also been argued that Joel Feinberg's open future argument provides a good reason to give parents the freedom to select against disabling conditions. Feinberg argues that parents hold rights in trust for their children (or future children) and it is possible for them to violate these rights, thus depriving their child of an open future. By using PGT, prospective parents are providing their child with a good start in life by ensuring that important options and experiences remain open to them. Therefore, prospective parents should have the freedom to select against any condition identified as disabling in order to give their child an open future.

This chapter has shown that there are good reasons for children not to begin their lives with unnecessary disadvantages and therefore good reasons for parents to have access to PGT for the purposes of selection against disabling conditions. As the arguments opposing selection against disabling conditions are flawed, there do not appear to be valid reasons for prohibiting parents from using PGT to select against disability. Lastly, I suggest that in constructing a regulatory framework that has national application, determinations as to seriousness should not play a part in the regulation of PGT.

Chapter 6
Testing Embryos for Polygenic Conditions and Traits

Abstract This chapter examines the ethical, scientific, and regulatory challenges of preimplantation genetic testing for polygenic conditions (PGT-P) and the use of this technology to select for non-medical traits. PGT-P allows testing for conditions influenced by multiple genes, including schizophrenia, diabetes, and heart disease. Unlike preimplantation genetic testing for monogenic conditions, which provides definitive results, PGT-P offers probabilistic risk scores, raising ethical concerns about environmental influences, disparities in predictive accuracy across ethnicities, and decision-making complexities due to pleiotropy. The chapter explores whether prospective parents should have the freedom to use PGT-P to test for polygenic conditions, and considers issues such as reproductive liberty, the potential to prevent debilitating conditions, and the importance of comprehensive genetic counselling. It also critically evaluates whether PGT-P should be used for trait selection, focusing on intelligence. Arguments raised in the previous chapter, such as procreative beneficence and the open future argument will be revisited, and new ethical objections introduced, including the therapeutic/non-therapeutic distinction and the 'playing God' argument. In conclusion, the chapter argues that while PGT-P poses unique challenges, it may be ethically permissible.

6.1 Introduction

As discussed in the preceding chapters, preimplantation genetic testing (PGT), specifically PGT for monogenic conditions (PGT-M), is currently used to test for and select against an array of single gene disorders, as well as for sex selection to avoid the inheritance of a sex-linked condition. Advances in our understanding of genetics are giving scientists the ability to predict the phenotype of particular combinations of genes with a greater degree of accuracy.[1] These scientific advances mean that it is now possible to test embryos for conditions controlled by multiple genes; these are

[1] Lello L, Raben TG, Hsu SDH (2020) Sibling Validation of Polygenic Scores and Complex Trait Prediction. Scientific Reports 10:13,190.

M. de Souza, *The Regulation of Embryo Testing in Australia*,
https://doi.org/10.1007/978-981-96-5902-9_6

known as polygenic conditions. This is made possible through the use of genome-wide association studies.[2] Statistical models called 'predictors' analyse the DNA of hundreds of thousands of individuals and are able to identify patterns in people's DNA.[3] Scientists can then pinpoint the patterns that correspond to various polygenic conditions. The advent of PGT for polygenic conditions (PGT-P) means that it is now possible to test embryos for conditions such as schizophrenia, diabetes[4] and heart disease. Whether it is acceptable to use PGT-P is complicated by the fact that whether or not the condition develops is also dependent on environmental factors.[5]

PGT-P can also be used to identify polygenic traits, for example, intelligence. In November 2018, news reports began to surface about scientists in the United States who had developed a test capable of predicting if an embryo would become a person of low intelligence.[6] Genomic Prediction, the biotechnology company which developed the test, claimed it could identify embryos that have a high chance of becoming a child with an intelligence quotient (IQ) of 25 points below average, and it is possible that in the future the same approach could be used to determine which embryos will develop into individuals of above average intelligence.[7] If the technology becomes available and safe, should parents have the freedom to test and select for IQ?

This chapter critically considers the use of PGT-P for identifying embryos that may develop into a person with a polygenic condition. It will also consider the use of PGT-P to select for polygenic traits, with a focus on intelligence. This chapter starts by asking how testing for polygenic conditions is different to the approved uses of PGT-M, PGT for structural rearrangements (PGT-SR) and PGT for aneuploidy (PGT-A). Consideration will then be given to the question of why parents should have the freedom to test for polygenic conditions. As polygenic conditions are disabling conditions, the arguments canvassed in Chap. 5 apply to polygenic conditions as well. The chapter then moves on to the issue of polygenic trait selection. As intelligence (or at least the part of intelligence that can be assessed through an IQ test) is one of only a small number of polygenic traits for which an embryo can be screened, I address the question of what is intelligence and why it might be a trait prospective parents

[2] National Human Genome Research Institute, "Genome-Wide Association Studies Fact Sheet.".

[3] Regolado A, Eugenics 2.0: We're at the Dawn of Choosing Embryos by Health, Height and More. *MIT Technology Review*, November 1, 2017. https://www.technologyreview.com/s/609204/eugenics-20-were-at-the-dawn-of-choosing-embryos-by-health-height-and-more/?source=download-metered-content.

[4] Treff NR, Eccles J, Lello L et al. (2019) Utility and First Clinical Application of Screening Embryos for Polygenic Disease Risk Reduction. Frontiers in Endocrinology 845. https://doi.org/10.3389/fendo.2019.00845.

[5] Siermann M, Tsuiko O, Vermeesch JR et al. (2022) A Review of Normative Documents on Preimplantation Genetic Testing: Recommendations for PGT-P. Genetics in Medicine 24: 1165–1175, 1165.

[6] Wilson C, 'A New Test Can Predict IVF Embryos' Risk of Having a Low IQ', *New Scientist* (online) 18 November 2018 < https://www.newscientist.com/article/mg24032041-900-exclusive-a-new-test-can-predict-ivf-embryos-risk-of-having-a-low-iq/ >.

[7] Ibid.

wish to select. I will re-consider two of the arguments put forward in the previous chapter: the principle of procreative beneficence[8] and the open future argument.[9] These arguments were considered persuasive as reasons to permit parents to select against disabling conditions. Are they as influential when considered in the context of selection for polygenic traits?

Following on from the previous chapter which argued that parents should have the freedom to select against disabling conditions, this chapter will explore the therapeutic/non-therapeutic distinction. This distinction is also known as the treatment–enhancement distinction and gives rise to the question, is there really a difference between allowing parents to avoid disabling conditions in their future offspring and permitting parents to choose the embryo that will become the most able child? If there is a difference, does it represent a good enough reason to prevent prospective parents from selecting for polygenic traits? I shall then re-examine the expressivist objection. This objection was discussed in detail in the previous chapter and therefore its application to positive selection need only be touched upon briefly. I will then consider the playing God argument. The chapter ends with the argument that prospective parents should have the freedom to use PGT-P to test for some polygenic traits.

6.2 Testing for Polygenic Conditions

PGT-P is a novel technology that has the potential to exacerbate the ethical tensions that already exist in the sphere of reproductive choice. It gives prospective parents the opportunity to test the embryos they have created for genetic conditions that are controlled by multiple genes. At present, it is possible to assess an embryo's risk of developing into a person with diabetes,[10] heart disease[11] or even schizophrenia.[12] The generation of a risk score is one of the reasons why PGT-P is controversial.[13] While PGT-M can give a definitive result about whether an embryo will develop into

---

[8] Julian Savulescu, 'Procreative Beneficence: Why We Should Select the Best Children' (2001) 15 *Bioethics* 413.

[9] Feinberg J (1980) 'The Child's Right to an Open Future' in William Aiken and Hugh LaFollette (eds), *Whose Child? Children's Rights, Parental Authority and State Power* (Littlefield, Adams & Co) 124.

[10] Treff NR, Eccles J, Lello L et al. (2019) Utility and First Clinical Application of Screening Embryos for Polygenic Disease Risk Reduction. Frontiers in Endocrinology 10: 1–6, 4.

[11] Treff NR, Eccles J, Marin D et al. (2020) Preimplantation Genetic Testing for Polygenic Disease Relative Risk Reduction: Evaluation of Genomic Index Performance in 11,883 Adult Sibling Pairs. *Genes* 11(6): 648.

[12] Tellier LCAM, Eccles J, Treff NR et al. (2021) Embryo Screening for Polygenic Disease Risk: Recent Advances and Ethical Considerations. Genes 12: 1106. See also Orchid Health, 'Identify Your Healthiest Embryo' https://www.orchidhealth.com Accessed 13 September 2023.

[13] See e.g., Professor Hank Greely, cited in Ball P 'Polygenic Screening of Embryos is Here, but is it Ethical?' The Guardian, 17 October 2021. https://www.theguardian.com/science/2021/oct/17/polygenic-screening-of-embryos-is-here-but-is-it-ethical Accessed 13 September 2023.

a person with a specific disabling condition,[14] PGT-P can only give a probabilistic indication. Therefore, it is possible that an embryo which has been rejected because of the risk that it may develop into a person with schizophrenia, if implanted would never have developed the condition.

The use of PGT-P is controversial for a number of reasons. First, as mentioned earlier, the development of polygenic conditions is partly dependent on environmental factors.[15] This means that if two embryos are assessed as being at the same risk of developing a condition, differences in their environment and lifestyle may mean that only one develops into an adult with the condition. Thus, embryos may be discarded even though may have developed into a healthy adult. A second concern is the is that the use of PGT-P to select against disabling conditions will lead to its use for polygenic traits. Whether this is a valid concern will be discussed later in this chapter. The third reason PGT-P is controversial is that the data collected in the genome wide association studies (from which risk scores are calculated), comprises individuals who are mainly of European ancestry.[16] This is problematic because it creates a disparity between the results generated for individuals of similar ancestry and those of other ethnic backgrounds, with more accurate results being generated for those of European heritage. For example, in the US type 2 diabetes carries a lifetime risk of 35.3%.[17] This means that the average American has a 35.3% chance of developing type 2 diabetes. The absolute risk[18] reduction that could be expected from PGT-P for a person of European ancestry is 5.5%, but for a person of African ancestry, PGT-P is only half as effective; 2.6%.[19] While this may be an issue now, Treff and Savulescu note that initiatives are underway across the world to increase the representation of individuals with non-European ancestry in biobanks.[20] Therefore, it will not necessarily be an issue in the near future as more data is collected and representation of other ethnicities is increased.

---

[14] It is worth noting that tests for lower penetrance single gene disorders (e.g. breast cancer) only indicate whether an embryo has the genetic mutations for that disorder. Such tests cannot say whether a person will actually develop, for example, breast cancer.

[15] Turley P, Meyer M, Wang N et al. (2021) Problems with Using Polygenic Scores to Select Embryos. N Engl J Med 385(1): 78–86, 79.

[16] Forzano F, Antonova O, Clarke A et al. (2021) The use of Polygenic Risk Scores in Pre-implantation Genetic Testing: An Unproven, Unethical Practice. European Journal of Human Genetics 30: 493–495, 493. https://doi.org/10.1038/s41431-021-01000-x.

[17] Turley P, Meyer M, Wang N et al. (2021) Problems with Using Polygenic Scores to Select Embryos. N Engl J Med 385(1): 78–86, 82.

[18] Absolute risk refers to the risk of (for our purposes) a disabling condition occurring in the general population.

[19] Turley P, Meyer M, Wang N et al. (2021) Problems with Using Polygenic Scores to Select Embryos. N Engl J Med 385(1): 78–86, 82.

[20] Treff NR, Savulescu J, de Melo-Martin I et al. (2022) Should Preimplantation Genetic Testing for Polygenic Disease be Offered to all—or none? Fertility and Sterility 117(6): 1162–1176, 1163.

A fourth point of contention is that PGT-P makes it possible to screen for multiple conditions at once.[21] Therefore, an embryo with a lower risk of developing a particular condition may be at an increased risk of developing another condition.[22] This is known as pleiotropy[23] and increases the complexity of the decision-making process for prospective parents.[24] Further, some scientists have argued that as the number of embryos created in an IVF cycle is often quite low, if embryos are to be screened for multiple conditions, it is possible that no embryos would be deemed suitable.[25] However, Nathan Treff and Julian Savulescu note that even if PGT-P is carried out on two embryos, evidence indicates a reduction in terms of polygenic disease risk compared to random selection.[26] They also argue that there is no evidence to support the contention that a lower risk of one condition may correlate with an increased risk of another; noting instead that risks are reduced in parallel.[27] On the other hand, Maria Siermann et al. note that the additional complexities presented by PGT-P, such as balancing the risks of multiple conditions, may mean that informed decision-making is difficult for prospective parents.[28]

A related concern noted by Siermann and her colleagues is that the differences between absolute and relative risk could confuse patients.[29] This is because while relative risk reduction might be high, the absolute risk reduction may actually be very

---

[21] Siermann M, Tsuiko O, Vermeesch JR et al. (2022) A Review of Normative Documents on Preimplantation Genetic Testing: Recommendations for PGT-P. Genetics in Medicine 24: 1165–1175, 1171.

[22] Forzano F, Antonova O, Clarke A et al. (2021) The use of Polygenic Risk Scores in Pre-implantation Genetic Testing: An Unproven, Unethical Practice. European Journal of Human Genetics 30: 493–495, 494. https://doi.org/10.1038/s41431-021-01000-x.

[23] Turley P, Meyer M, Wang N et al. (2021)Problems with Using Polygenic Scores to Select Embryos. N Engl J Med 385(1): 78–86, 80.

[24] An alternative problem identified by Forzano et al. is that prospective parents will only be informed of the risk of the condition of which they are aware and not of the risk of a condition they have not sought information about— Forzano F, Antonova O, Clarke A et al. (2021) The use of Polygenic Risk ScoresPolygenic risk score in Pre-implantation Genetic Testing: An Unproven, Unethical Practice. European Journal of Human Genetics 30: 493–495, 494. https://doi.org/10.1038/s41431-021-01000-x .

[25] Ibid.

[26] Treff NR, Savulescu J, de Melo-Martin I et al. (2022) Should Preimplantation Genetic Testing for Polygenic Disease be Offered to all—or none? Fertility and Sterility 117(6): 1162–1176, 1163.

[27] Ibid.

[28] Siermann M, Tsuiko O, Vermeesch JR et al. (2022) A Review of Normative Documents on Preimplantation Genetic Testing: Recommendations for PGT-P. Genetics in Medicine 24: 1165–1175, 1172. See also Komorowski AS, Feinberg EC (2022) Scientific and Ethical Considerations in Using Preimplantation Genetic Testing for Polygenic Disease. Fertility and Sterility 117(6): 1160–1161, 1160.

[29] Siermann M, Tsuiko O, Vermeesch JR et al. (2022) A Review of Normative Documents on Preimplantation Genetic Testing: Recommendations for PGT-P. Genetics in Medicine 24: 1165–1175, 1166.

low.[30] Conversely, Treff and Savulescu argue that 'it is unethically paternalistic to believe' that if given adequate information individuals would be unable to understand these aspects of PGT-P and make their own decisions.[31] Siermann et al., however, argue that it is not only prospective parents who struggle to understand the complexities of PGT-P, some healthcare providers do as well.[32] Rather than considering this as a reason to prevent the use of PGT-P, it should instead highlight the need for all patients to receive comprehensive genetic counselling and for genetic counsellors to receive adequate training.

The importance of comprehensive genetic counselling can also be evinced in the context of polygenic conditions that fall on a continuum of severity. Turley and his colleagues illustrate this point in the context of idiopathic short stature, which they define as 'a height that is more than 2 standard deviations below the mean'.[33] Thus an embryo may be considered at risk of developing into a person of short stature and discarded, when that embryo may have developed into a person only 2.5 cm shorter than the embryo that was ultimately selected.[34] It is therefore important that prospective parents are counselled to ensure that they understand that reducing the risk of certain conditions through PGT-P may not produce a significant difference in the phenotype of the future child.[35] While Turley's argument may carry weight in the context of disabling conditions such as intellectual disability, it is worth noting that data from genome wide association studies can predict height to within a few centimetres.[36] It may therefore be possible for prospective parents not only to select against idiopathic short stature, but to select an embryo which will develop into an individual of a specific height. Nevertheless, ensuring prospective parents have a realistic idea of what PGT-P can achieve, highlights not only the importance of genetic counselling but also the necessity of educating the public about the capabilities of new developments in PGT.

---

[30] Consider Type 1 diabetes. 0.21% of individuals in Australia will develop Type 1 diabetes by the age of 19: https://www.aihw.gov.au/reports/diabetes/diabetes/contents/how-common-is-diabetes/type-1-diabetes. If PGT-P could halve the risk of selecting an embryo with Type 1 diabetes, the relative risk reduction of PGT-P could be advertised as 50%. This would be confusing for prospective parents because the absolute risk reduction is half of 0.21%, i.e. 0.1%.

[31] Treff NR, Savulescu J, de Melo-Martin I et al. (2022) Should Preimplantation Genetic Testing for Polygenic Disease be Offered to all—or none? Fertility and Sterility 117(6): 1162–1176, 1165.

[32] Siermann M, Vlacke O, Vermeesch JR et al. (2023) Limitations, Concerns and Potential: Attitudes of Healthcare Professionals towards Preimplantation Genetic Testing Using Polygenic Risk Scores. European Journal of Human Genetics 31:1133–1138, 1137.

[33] Turley P, Meyer M, Wang N et al. (2021) Problems with Using Polygenic Scores to Select Embryos. N Engl J Med 385(1): 78–86, 81.

[34] Ibid.

[35] Turley et al. (ibid) refer to such an outcome as not 'practically meaningful', however this is not necessarily the case, as 2.5 cm could be the difference between an embryo developing into a person who meets the height requirements for certain jobs, e.g. a pilot, and someone who does not.

[36] Lello L, Avery SG, Tellier L et al. (2018) Accurate Genomic Prediction of Human Height. Genetics 210: 477–497. https://doi.org/10.1534/genetics.118.301267.

A further reason that PGT-P is controversial is because while polygenic risk scores[37] have clinical utility in the context of understanding disease mechanisms in existing individuals, its usefulness at the embryonic stage as a selection tool is currently unproven.[38] As Inmaculada de Melo Martin and Carl Shulman note, the accuracy of PGT-P will not become apparent until decades after the individual is born.[39] However, if PGT-P is not permitted, its accuracy, or otherwise, will never be demonstrated.

In deciding whether a revised regulatory framework should permit PGT-P, the reproductive liberty of prospective parents also needs to be considered. As the biopsy process for PGT-P is the same as for the approved uses of PGT, the physical risk to the embryo is no different. Therefore, reproductive liberty cannot be restricted on the basis of physical harm to the embryo. PGT-P can indicate the risk of an embryo developing into a person with a condition such as schizophrenia that will affect them for a substantial part of their life and require lifelong treatment[40]; therefore it seems beneficial for prospective parents to have the freedom to use PGT-P. Reducing the chance of a future child developing a debilitating condition will also benefit the prospective parents, who will not have to cope with the premature death of their child from a condition that could have been screened out.[41] A study run by Jennifer Eccles and her colleagues found that when PGT-P was offered (at no cost) to prospective parents who were already using PGT-A, over 50% of patients elected to receive a polygenic risk score for their embryos.[42] Further, when given a choice between receiving a polygenic risk score for all twelve of the disabling conditions offered, or a selection thereof, 100% of the patients opting for PGT-P chose to screen for all conditions.[43] This suggests that if given a choice, the majority of prospective parents undergoing PGT for an already approved purpose would also choose to use PGT-P.

[37] In the context of embryo testing, a polygenic risk score indicates the likelihood of an embryo developing into a person with a polygenic condition or trait, based on an estimate of its genetic component.

[38] Treff NR, Savulescu J, de Melo-Martin I et al. (2022) Should Preimplantation Genetic Testing for Polygenic Disease be Offered to all—or none? Fertility and Sterility 117(6): 1162–1176, 1163. NB: This article is part of Fertility and Sterility's 'Fertile Battle' series, where a group of authors argue different sides of a contentious proposition. Savulescu and Treff argued in favour of the proposition detailed in the title of the article, while de Melo-Martin and Shulman argued against it.

[39] Ibid. Concerns about the lack of scientific knowledge around polygenic risk scores in the context of embryo selection were also voiced by reproductive healthcare professionals in a study by Siermann et al.: Siermann M, Vlacke O, Vermeesch JR et al. (2023) Limitations, Concerns and Potential: Attitudes of Healthcare Professionals towards Preimplantation Genetic Testing Using Polygenic Risk Scores. European Journal of Human Genetics 31:1133–1138, 1134.

[40] Andreasen, NC (2011) Concept of Schizophrenia: Past, Present and Future. In: Weinberger DR, Harrison PJ (eds) Schizophrenia, 3rd edn. Wiley-Blackwell, Chichester, pp 3–8; Seifert R (2019) Basic Knowledge of Pharmacology. Springer, Berlin, 346.

[41] Treff NR, Savulescu J, de Melo-Martin I et al. (2022) Should Preimplantation Genetic Testing for Polygenic Disease be Offered to all—or none? Fertility and Sterility 117(6): 1162–1176, 1164.

[42] Eccles J, Marin D, Duffy L et al. (2021) Rate of Patients Electing for Polygenic Risk Scores in Preimplantation Genetic Testing. Fertility and Sterility 116(3) Supplement: e267–e268.

[43] Ibid.

If PGT-P were to play a role in a regulatory framework for embryo testing, would a review of PGT-M documents provide any assistance? Siermann et al. conducted a large scale review of documents and guidelines involved in the regulation of PGT-M, with the aim of identifying the similarities and differences that would be relevant for developing guidance on PGT-P.[44] They discovered certain themes across these documents including those related to the circumstances in which PGT-M is considered appropriate.[45] Many of these documents consider PGT-M appropriate for serious conditions or conditions that cause suffering. However, in Chap. 5 I argued that the criterion of seriousness should not form part of a revised framework for the regulation of PGT. Further, the documents examined by Siermann's team date back as far as 1996, with many from the early 2000s,[46] perhaps indicative of a more cautious approach than is needed today. The documents also indicated disapproval for using PGT to test for non-medical traits,[47] which at the time many of these documents were written was not even possible. Some documents referred to polygenic conditions, however, there was disagreement about whether or not PGT was acceptable for such conditions.[48] The documents expressing disapproval for PGT-P were over a decade old,[49] suggesting perhaps a need for review in light of the scientific advances of the last ten years.[50]

One thing that is clear from Siermann's review of PGT-M guidance is that what are considered appropriate uses of PGT-M differs across countries, even though the technology used in PGT-M is well established. In light of this, it is unsurprising that there is anxiety around the much newer technology of PGT-P. It should also be borne in mind that PGT-P differs from PGT-M, because rather than testing indicating that an embryo has a particular disabling condition and should be discarded, PGT-P provides a risk score that can be used to rank embryos.[51] Accordingly, Siermann and her team acknowledge that this distinction may mean that the regulatory framework

[44] Siermann M, Tsuiko O, Vermeesch JR et al. (2022) A Review of Normative Documents on Preimplantation Genetic Testing: Recommendations for PGT-P. Genetics in Medicine 24: 1165–1175.

[45] Siermann M, Tsuiko O, Vermeesch JR et al. (2022) A Review of Normative Documents on Preimplantation Genetic Testing: Recommendations for PGT-P. Genetics in Medicine 24: 1165–1175, 1167.

[46] Ibid 1168–1169.

[47] Ibid 1169.

[48] Ibid 1170.

[49] See, e.g. Commission for Ethics in Science and Technology. Position statement ethics and assisted procreation: guidelines for the donation of gametes and embryos, surrogacy and preimplantation genetic diagnosis. Published 30 September 2009. https://www.ethique.gouv.qc.ca/wp-content/uploads/ldmorh3e_paresume-et-errata-en.pdf. Accessed 4 August 2025.

[50] The review also touched upon the ethical acceptability of PGT-M for carrier screening. Once again, different views about whether carrier screening is an acceptable use of PGT were expressed: Siermann M, Tsuiko O, Vermeesch JR et al. (2022) A Review of Normative Documents on Preimplantation Genetic Testing: Recommendations for PGT-P. Genetics in Medicine 24: 1165–1175, 1170.

[51] Treff NR, Savulescu J, de Melo-Martin I et al. (2022) Should Preimplantation Genetic Testing for Polygenic Disease be Offered to all—or none? Fertility and Sterility 117(6): 1162–1176, 1163.

for PGT-M is an unsuitable springboard for the development of guidance around the use of PGT-P.[52] Siermann concludes that 'ethical consensus for PGT-P is absent, and potential support will at most be conditional'.[53] While this may be a sound conclusion based on the documentation under consideration, it is arguable whether documentation that is not focused on PGT-P is even a suitable basis on which to found such a conclusion.

My aim in this chapter is to ascertain whether prospective parents should have the freedom to use PGT-P and if so, for what purposes. The literature on PGT-P suggests that many of the concerns around its use stem from the novelty of the technology.[54] As Meyer et al. point out, there was a similar level of apprehension during the early days of IVF. Citing a survey by market research firm The Harris Poll, Meyer and her colleagues note that in 1969 most Americans found IVF objectionable.[55] However, nine years later, and only a month after the birth of Louise Brown, the same poll showed a marked shift in public opinion, with over 60% of the Americans surveyed indicating not only that they did not have objections to the use of IVF, but also that they would consider using it themselves.[56] Thus Meyer and her colleagues conclude that 'innovations that are initially met with … active resistance can quickly become normalized and widely adopted'.[57] With greater public engagement and proper regulatory oversight,[58] testing embryos for polygenic conditions could become an acceptable practice. Therefore, the use of PGT-P for disabling conditions should feature in the development of a revised regulatory framework for embryo testing. In constructing this framework, it is important to consider novel uses of PGT and how appropriate regulatory oversight can be achieved. This will be discussed further in Chap. 8. In the next section, I turn to the question of whether prospective parents should be permitted to use PGT-P to test for non-medical traits.

---

[52] Siermann M, Tsuiko O, Vermeesch JR et al. (2022) A Review of Normative Documents on Preimplantation Genetic Testing: Recommendations for PGT-P. Genetics in Medicine 24: 1165–1175, 1171.

[53] Siermann M, Tsuiko O, Vermeesch JR et al. (2022) A Review of Normative Documents on Preimplantation Genetic Testing: Recommendations for PGT-P. Genetics in Medicine 24: 1165–1175, 1171.

[54] See, eg, Siermann M, Vlacke O, Vermeesch JR et al. (2023) Limitations, Concerns and Potential: Attitudes of Healthcare Professionals towards Preimplantation Genetic Testing Using Polygenic Risk Scores. European Journal of Human Genetics 31:1133–1138, 1136.

[55] Meyer MN, Tan T, Benjamin DJ, et al. (2023) Public Views on Polygenic Screening of Embryos: Understanding Moral Acceptability and Willingness to Use is Crucial for Informing Policy. Science 379(6632): 541–543, 543.

[56] Ibid.

[57] Meyer MN, Tan T, Benjamin DJ, et al. (2023) Public Views on Polygenic Screening of Embryos: Understanding Moral Acceptability and Willingness to Use is Crucial for Informing Policy. Science 379(6632): 541–543, 543.

[58] Forzano F, Antonova O, Clarke A et al. (2021) The use of Polygenic Risk Scores in Preimplantation Genetic Testing: An Unproven, Unethical Practice. European Journal of Human Genetics 30: 493–495, 494. https://doi.org/10.1038/s41431-021-01000-x.

6.3 Testing Embryos for Polygenic Traits

One of the concerns around the use of PGT-P is that it may be used for the selection of polygenic traits. While such tests are not yet commercially available, the technology to test for certain complex traits exists[59] and thus it may only be a matter of time before this question needs to be confronted by regulatory frameworks across Australia. A US study into public attitudes to PGT-P indicated that over half of the research participants had no moral objection to its use for both medical and non-medical purposes.[60] However, when asked if they would use PGT-P to increase the chance of their future child attending a top university, this figure dropped to 43%.[61] The study also showed that of those under 35, 48% were open to using PGT-P to improve their future child's educational performance.[62] Further, regardless of age, 48% of participants with a university education were amenable to the use of PGT-P to increase their future child's chance of attending a highly ranked university. This indicates significant support (at least in the US) for the use of PGT-P for non-medical traits. It also illustrates that the target population for such testing (i.e. those of reproductive age), are more receptive to its use.

The arguments discussed in the context of using PGT-P to select against polygenic conditions also apply when thinking about non-medical uses of PGT-P. Consider again pleiotropy. Pleiotropy may be defined as 'the tendency of genetic variants to affect multiple phenotypes.'[63] For example, Turley et al. note that embryos with high polygenic scores for educational attainment, also showed an increased risk of bipolar disorder.[64] To ameliorate this pleiotropic effect, prospective parents may wish to select an embryo with a high polygenic score for educational attainment that also has a low polygenic score for bipolar disorder. This may mean that the embryo with the highest polygenic score for educational attainment is not selected, thus reducing the benefit that may be expected from using PGT-P.[65]

This section focuses on whether prospective parents should have the freedom to use PGT-P to test for non-medical traits that may prove advantageous to the future child. Such traits may be termed 'abilities'; but just like disability, it is almost impossible to develop a coherent theory of what counts as an ability in all contexts. This is because abilities are context dependent. This can be illustrated by looking at

---

[59] Wilson C, 'A New Test Can Predict IVF Embryos' Risk of Having a Low IQ', *New Scientist* (online) 18 November 2018 < https://www.newscientist.com/article/mg24032041-900-exclusive-a-new-test-can-predict-ivf-embryos-risk-of-having-a-low-iq/ > .

[60] Meyer MN, Tan T, Benjamin DJ, et al. (2023) Public Views on Polygenic Screening of Embryos: Understanding Moral Acceptability and Willingness to Use is Crucial for Informing Policy. Science 379(6632): 541–543, 542.

[61] Ibid.

[62] Ibid.

[63] Turley P, Meyer M, Wang N et al. (2021) Problems with Using Polygenic Scores to Select Embryos. N Engl J Med 385(1): 78–86, 80.

[64] Ibid 81.

[65] Ibid.

the dictionary definition of ability. For example, the Oxford Dictionary of English defines an ability as 'possession of the means or skill to do something'.[66] Thus a particular skill only qualifies as an ability if the person possessing that skill needs (or wants) to do the task to which the skill relates. Therefore, if a child with a range of abilities which would enable them to thrive in context B grows up in context A, they may not flourish. Despite the difficulties inherent in formulating a conception of ability capable of application in the context of reproductive choice, attempts have been made by some philosophers. For example, Allen Buchanan et al. have introduced the idea of 'general-purpose means': traits that would prove valuable to a person, no matter what sort of life they choose to lead.[67] This is based on John Rawls' concept of natural primary goods: things that 'normally have a use whatever a person's rational plan of life'.[68]

General-purpose means may include 'enhanced immune capabilities', improved memory and better powers of concentration.[69] Savulescu suggests intelligence, self-discipline, impulse control, foresight, patience, sense of humour, and a sunny temperament may also fall within this category.[70] While the value placed on each of these traits may differ depending on the particular life choices a person makes, general-purpose means would prove valuable in any context.[71] Consider enhanced immune capabilities. A person with a stronger immune system would be protected from illnesses that may severely debilitate others[72]; this would have been particularly beneficial during the COVID-19 pandemic. Whatever course of life the person chooses to pursue, not contracting illnesses (or contracting significantly fewer than the average person) will be beneficial to their quality of life.

Certain abilities are only advantageous in so far as they confer a competitive advantage on the person who has that ability.[73] However, if large sections of society are all in possession of that ability, the competitive edge disappears. For example, if everyone could run 100 m in 9.58 s, Usain Bolt, the current world record holder, would not be regarded as the fastest man on earth. It is, however, worth bearing in mind that parents utilising PGT-P only have the option to select from the embryos they have created. Therefore, while some parents may wish to select for athletic ability, none of the embryos they create may have the genetic makeup to make them the next Usain Bolt. However, parents may wish to select the embryo which will become the

---

[66] *Oxford Dictionary of English*, 3rd edn (online).

[67] Buchanan A et al. (2000) From Chance to Choice: Genetics and Justice. Cambridge University Press, 167.

[68] Rawls J, *A Theory of Justice* (Oxford University Press, 1971) 62.

[69] Buchanan A et al., *From Chance to Choice: Genetics and Justice* (Cambridge University Press, 2000) 186.

[70] Savulescu J (2007) In Defence of Procreative Beneficence. Journal of Medical Ethics 33: 284, 284.

[71] Buchanan A et al., From Chance to Choice: Genetics and Justice (Cambridge University Press, 2000) 168.

[72] E.g. viruses such as COVID-19 or the flu, or bacterial infections such as meningitis or pneumonia.

[73] Buchanan A et al., From Chance to Choice: Genetics and Justice (Cambridge University Press, 2000), 184.

child with more athleticism than the other possible children, and this could still be advantageous to the child because fitness, strength and agility are perceived to be desirable characteristics to possess. Accordingly, a trait like athleticism which may be considered a 'positional trait' may also have intrinsic value.[74]

For general-purpose means to form part of the regulatory framework for PGT-P, consideration needs to be given to who decides which traits count as a general-purpose means. In Chap. 4, I suggested that the interactional model of disability would facilitate decision-making by enabling prospective parents and medical professionals to work together to decide if a genetic condition is actually a disabling condition. In a similar vein, a joint decision-making process between medical professionals and prospective parents could be undertaken to decide which traits would prove valuable to the future child, given the environment into which they will be born. The medical professional will be in a position to identify which combinations of traits are present in the embryo, as well as to advise on matters such as the role played by environmental factors. Prospective parents, on the other hand, are best placed to evaluate the future child's social environment. Therefore, decisions regarding which traits fall within the category of general-purpose means are not objective decisions to be arrived at through the application of rigid rules.

Of the polygenic traits for which PGT-P may be utilised in the future, IQ is one of the few non-medical traits for which an embryo can currently be tested. As IQ is a measure of intelligence, this chapter will now turn to consider what intelligence is and why it may be classed as a trait that would generally be valuable to a person. It then moves on to consider whether conclusions about the permissibility of testing embryos for IQ can be applied to other general-purpose means, to answer the question, should prospective parents have the freedom to test their embryos for traits that may be classed as abilities? To answer this question, two arguments that were introduced in the previous chapter will be revisited; procreative beneficence and the open future argument. I then analyse the therapeutic–non-therapeutic distinction and the expressivist objection. Following on from this, I turn to the Playing God argument.

6.3.1 The Case of IQ

> Being intelligent matters; it makes a big difference to human lives. Sharp men and women thrive. Problems are solved by astute thinkers. Questions get answered by those who are clever.[75]

---

[74] Ibid 186.

[75] Howe M (1997) IQ in Question. SAGE, London, 1.

Intelligence may be described in a number of ways, including, 'the ability to acquire and apply knowledge and skills'[76]; a person's overall capacity to 'act purposefully, to think rationally, and to deal effectively with his environment',[77] and as 'innate general cognitive ability'.[78] A broader description is given by Howard Gardner, who suggests that the word 'intelligence' may be used in three distinct ways.[79] First, as a 'property of all human beings'.[80] Second, as a 'dimension on which human beings differ', and third, to describe the manner in which a task is carried out in pursuit of a goal.[81] Gardner also notes that there are different kinds of intelligence, such as linguistic intelligence,[82] musical intelligence[83] and 'logical-mathematical intelligence',[84] to name a few. While only a small selection of possible definitions are outlined above, the theme running through them is that intelligence is linked to ability and capacity.

As noted above, IQ is a measure of intelligence, and intelligence as a concept is broad. It is therefore worth touching upon which aspects of intelligence IQ tests measure. Such tests generally measure verbal communication, mathematical skills, spatial awareness and memory.[85] These represent broader abilities such as problem-solving skills, comprehension, processing speed and visual-spatial thinking.[86] As not all kinds of intelligence are measured by IQ tests, using genome-wide association studies to predict IQ is problematic. For example, as musical intelligence is not measured by IQ tests, an embryo that may become a person with a high level of musical intelligence may not be recognised as such via a genome-wide association study that uses IQ as a basis for predicting intelligence.[87]

IQ tests have stirred controversy due to their perceived bias towards the middle class and certain races.[88] However, according to Marks, what was traditionally interpreted as a race bias may be explained by differences between native and non-native

---

[76] *Oxford Dictionary of English* (Oxford University Press, 2010) 2324.

[77] David Wechsler, quoted in Bartholomew DJ, *Measuring Intelligence: Facts and Fallacies* (Cambridge University Press, 2004) 4.

[78] Cyril Burt, quoted ibid.

[79] Howard Gardner, *Frames of Mind: The Theory of Multiple Intelligences* (Basic Books, 2011) xv.

[80] Ibid.

[81] Ibid.

[82] Ibid 122.

[83] Gardner H (2011) Frames of Mind: The Theory of Multiple Intelligences. Basic Books, 148.

[84] Gardner, Frames of Mind, 176.

[85] See, e.g., the Woodcock-Johnson Tests for Cognitive Abilities.

[86] Taub GE, McGrew KS (2004) A Confirmatory Factor Analysis of Cattell-Horn-Carroll Theory and Cross-Age Invariance of the Woodcock-Johnson Tests of Cognitive Abilities III. School Psychology Quarterly 19(1): 72–87. http://dx.doi.org/10.1521/scpq.19.1.72.2940974.

[87] There may also be issues with small sample sizes for genome-wide association studies as not everyone has taken an IQ test. Therefore, while genome-wide association study sample sizes of 280,000 exist for IQ, it may be difficult to reach one million.

[88] See, e.g., Castles E (2012) Inventing Intelligence: How Americans Came to Worship IQ. Praeger, Santa Barbara, 92. Castles notes that scholastic aptitude test (SAT) scores (which correlate very highly with IQ test scores) show a correlation with parental income that is similar to that of SAT

English speakers.[89] Consequently, if a test taker struggles to understand the instructions, their ability to answer the questions will be compromised.[90] To overcome this difficulty, modern IQ tests have been developed to be more culturally sensitive than their predecessors.[91]

However, overcoming cultural bias is in itself a difficult task, as 'all tests are to some degree culturally loaded to the extent that they reflect information that is relevant and valued in a cultural context'.[92] Therefore, as much as possible, contemporary assessments of IQ have been developed with an understanding of the effect of culture on testing.[93] For example, Margaret Lanca and Emily Wilner note that for testing to be carried out successfully, the participant must understand the concept of testing, meaning that they are aware that they need to perform at their best.[94] However, test takers from cultures that do not value formal assessment may be unfamiliar with this expectation and may find it uncomfortable.[95] Similarly, some tests must be completed within a specific time frame; but the necessity to complete a task quickly may also be culturally dependent, with some cultures perceiving speed and quality to be at odds with each other.[96] Therefore, contemporary IQ tests take into account that an interpreter may be involved in the process and also utilise 'multicultural assessment tools'.[97] However, it is unclear whether the individuals whose data comprises current genome-wide association studies had access to these multicultural assessments and, therefore, whether the IQ scores represented in such data are devoid of cultural biases. Further, IQ testing is not a standard part of education in many countries and thus individuals whose genomic data has an associated IQ score may predominantly come from specific geographical locations. As environmental factors play a role in the expression of a polygenic trait, an embryo selected on the basis that it will develop into a person of high IQ may not achieve this due to the impact of a different educational environment.

Despite the problems inherent in IQ testing, intelligence may still be a trait for which prospective parents wish to select. Studies have shown that intelligence,

---

scores and undergraduate grades. She thus notes that parental income may be as good an indicator of university success as test scores.

89 Marks, "IQ Variations," 643.

90 Castles, "Inventing Intelligence," 124.

91 Lanca M, Wilner EK (2019) Neuropsychology in the Twenty-First Century: The Rise of Multicultural Assessments. In: Sanders KM (ed) Physician's Field Guide to Neuropsychology. Springer, New York, 479.

92 Suzuki LA, Naqvi S, Hill JS (2014) Assessing Intelligence in a Cultural Context. In: Leong FTL (ed) APA Handbook of Multicultural Psychology, vol 1. American Psychological Association, Washington: p 247–266, 260.

93 Lanca M, Wilner EK (2019) Neuropsychology in the Twenty-First Century: The Rise of Multicultural Assessments. In: Sanders KM (ed) Physician's Field Guide to Neuropsychology. Springer, New York, 479.

94 Ibid 481.

95 Ibid.

96 Ibid.

97 Ibid, 480.

together with certain 'temperamental traits', can have a positive effect on diagnoses of certain mental illnesses,[98] including major depression and schizophrenia.[99] Cederblad and her colleagues suggest that this is because individuals who have a high level of intelligence are better able to evaluate their problems accurately and utilise appropriate coping mechanisms.[100] As well as enabling individuals to cope with challenging situations, intelligence also has positive effects on other aspects of a person's life. Plomin remarks that 'life is an intelligence test' because intelligence affects everything from school performance and social opportunities to even the romantic choices a person makes.[101] Further, he notes that intelligence predicts educational, employment and 'health outcomes better than any other trait'.[102] This suggests that moderate and high levels of intelligence may be beneficial to a future child in various ways, whereas low levels of intelligence may be associated with adverse outcomes such as poor school performance and fewer social opportunities. Consequently, applying such findings to the use of polygenic screening to test for IQ, it follows that prospective parents should have the freedom to exercise their reproductive liberty and utilise such tests once they become available in Australia.

Having concluded that prospective parents should be permitted to use PGT-P to select for IQ, this chapter turns to consider whether it is possible to identify broader principles that should govern decision-making in this area.

6.3.2 Procreative Beneficence

In Chap. 5, Julian Savulescu's principle of procreative beneficence was introduced. It may be recalled that procreative beneficence is the idea that prospective parents 'have a significant moral reason to select the child, of the possible children they could have,

---

[98] Cederblad M et al. (1995) Intelligence and Temperament as Protective Factors for Mental Health: A Cross-sectional and Prospective Epidemiological study. European Archives of Psychiatry and Clinical Neuroscience 245: 11, 16.

[99] Gale CR et al., 'Is Bipolar Disorder More Common in Highly Intelligent People? A Cohort Study of a Million Men' (2013) 18 *Molecular Psychiatry* 190, 190. See also Mortensen EL et al. (2005) IQ and Mental Disorder in Young Men. 187 *British Journal of Psychiatry* 407. Mortensen et al. found low levels of intelligence were more commonly associated with schizophrenia and other psychotic disorders. Similar results were reported in Hill WD, Harris SE, Deary IJ (2019) What Genome-wide Association Studies Reveal About the Association Between Intelligence and Mental Health. *Current Opinion in Psychology* 27: 25.

[100] Cederblad M et al. (1995) Intelligence and Temperament as Protective Factors for Mental Health: A Cross-sectional and Prospective Epidemiological study. European Archives of Psychiatry and Clinical Neuroscience 245: 11, 18.

[101] Plomin R, von Stum S "The New Genetics of Intelligence," Nature Reviews: Genetics 19, no 3 (2018): 148–159, 148.
https://doi.org/10.1038/nrg.2017.104.

[102] Ibid.

whose life is expected… to go best or at least not worse than any of the others'.[103] It was noted in the previous chapter that one of the criticisms of procreative beneficence stems from the idea of the 'best life'. In the context of disabling conditions, the best life may be considered to be the life with the fewest limitations. For example, a life limited by severe chronic pain may arguably not be the best life. Deciding what may be considered the best life adds to an already complex reproductive decision-making process for prospective parents. However, if parents are to be given the freedom to use PGT-P to select for traits that may be classed as abilities, this issue needs to be addressed.

Buchanan et al ask 'whether there are adequate or defensible standards for determining what would be the best children possible'.[104] Further, the authors state that it

> is important that the attempts to produce the best child possible be understood as making the life of the child best for the child from the standpoint of that particular child's good, not best from some other standpoint, such as the good of the parents or the good of society.[105]

This leads to questions such as, would it better for a child to have above average mathematical skills but poor social skills, or would a child be better equipped with average mathematical skills and excellent interpersonal skills? Savulescu's answer is that procreative beneficence can be 'sufficiently determinative' without precisely ranking each possible embryo.[106] Instead, a broad-brush approach to rankings may be possible. For example, it 'may not be possible to say whether A is better than B, but it may be possible to say that both A or B are better than C. This is enough to rationally reject C'.[107] Thus Savulescu notes that certain sequences of traits may be 'inferior to others, and that is enough for procreative beneficence to be of value'.[108] This highlights the value of procreative beneficence in the context of negative selection, where it can focus on which disabling conditions should be avoided. While it cannot provide definitive answers to the quandary of which traits should be valued over others, the principle of procreative beneficence may still play a valuable role in a regulatory framework for reproductive choice.

Rob Sparrow argues that the idea of the best child is implausible, regardless of the perspective from which the best life is judged. He states that a hedonistic viewpoint would be a useless perspective from which to judge which future child will have the best life because a 'good life is more than a merely pleasurable life'.[109] Using an

---

[103] Savulescu J, Kahane G, 'The Moral Obligation to Create Children with the Best Chance of the Best Life' (2009) 23 *Bioethics* 274, 274.

[104] Buchanan A et al. (2000) From Chance to Choice: Genetics and Justice. Cambridge University Press, 161.

[105] Ibid 164.

[106] Savulescu J (2007) In Defence of Procreative Beneficence. J Med Ethics 33: 284–288, 285.

[107] Ibid.

[108] Ibid.

[109] Sparrow R (2007) Procreative Beneficence, Obligation, and Eugenics. Genomics, Society and Policy 3(3): 43–59, 48.

objective list theory[110] as a starting point, Sparrow suggests that a good life could consist of objective goods such as love and happiness and 'it may well be possible to choose children who are more likely to succeed in the pursuit of these goods'.[111] However, the problem of identifying the best life still remains. Sparrow notes that

> Any plausible account of objective goods will include goods of widely different natures, which are probably incommensurable. Any account of the relative merits of these competing goods and the sorts of trade-offs we should make between them will be a thousand times more controversial than this list of goods itself.[112]

Thus Sparrow concludes that choosing the best life based on an objective list theory 'is at best extremely controversial, if not ultimately impossible'.[113] He does, however, note that there may well be reasons to select embryos that possess 'genes for enhanced capacities'.[114] Similarly, Michael Parker rejects pursuit of the best possible child, but notes the importance of beneficence as a 'moral dimension of reproductive choice'[115] and therefore supports testing for and selecting in favour of 'conditions for the possibility of a good life'.[116] Further, Parker asserts that what makes a life go well is made up of 'complex and interdependent concepts' which will vary not just from community to community, but also from family to family.[117]

This discussion has shown that while the principle of procreative beneficence is not without its problems, it has the potential to be an important principle in the regulatory framework for embryo testing, both in the context of selection against disabling conditions and in the context of selection for abilities. As the idea of the best life is quixotic, it should be abandoned in favour of a more modest goal. To realise its regulatory potential, it is suggested that perhaps the principle of procreative beneficence should read as follows: parents have moral reasons 'to select the child, of the possible children they could have'[118] who will have a range of 'general purpose means' that are conducive to a life going well.[119] If more than one embryo can be expected to become a child who will have this range of traits, parents act morally by choosing any of these embryos.

---

[110] Objective list theories of wellbeing suggest that there are a number of basic goods that can objectively be judged as being beneficial to people. See generally, Rice CM (2013) Defending the Objective List Theory of Well-Being. Ratio 26(2): 196.

[111] Sparrow R (2007) Procreative Beneficence, Obligation, and Eugenics. Genomics, Society and Policy 3(3): 43–59, 49.

[112] Ibid.

[113] Ibid 50.

[114] Ibid.

[115] Parker M (2007) The Best Possible Child. 33 *Journal of Medical Ethics* 279, 282.

[116] Ibid.

[117] Ibid.

[118] Savulescu J, Kahane G (2009) The Moral Obligation to Create Children with the Best Chance of the Best Life. 23 *Bioethics* 274, 274.

[119] Bartels suggests that genetics account for 35% of wellbeing; Bartels M (2015) Genetics of Wellbeing and its Components Satisfaction with Life, Happiness, and Quality of Life: A Review and Meta-Analysis of Heritability Studies. Behavioural Genetics 45: 137–156, 154.

6.3.3 The Open Future Argument

This section re-considers Joel Feinberg's Open Future argument, which was discussed in Chap. 5. This is the argument that prospective parents should not violate the 'rights in trust' they hold for their children.[120] If they do, they may prematurely close off key future options that child has, thereby denying them the right to an open future. In Chap. 5, I suggested that Feinberg's argument provided a good basis for permitting parents to use PGT-M to select against disabling conditions. Is the open future argument as persuasive in the context of PGT-P for traits that may be classed as abilities?

Rob Sparrow argues that rather than trying to provide a child with an open future, prospective parents should aim to give their child 'a reasonable range of valuable life-options'.[121] He refers to this as 'a child's right to a *decent* future'.[122] To comprise a decent future, two conditions must be met. The first is that 'the child must have a reasonable range of options available to them in order to ensure that they have a chance of flourishing'.[123] Second, the options must predominantly comprise choices that could make up 'a worthwhile human life'.[124] Sparrow asserts that a decent future differs from an open future because for a child to have a decent future they do not need to be provided with 'as many options as possible'.[125] It is, however, unclear what Sparrow means by a 'reasonable range of options' and why this is preferable to Feinberg's argument that a 'child's future [be] left open as much as possible for his own finished self to determine'.[126] If prospective parents undergo IVF with PGT and discover that all the embryos are healthy, is this enough to satisfy Sparrow's 'decent future' argument?[127] Does this therefore mean that prospective parents should not have the freedom to test for abilities? For Sparrow, what is important is that parents are not *required* to provide their future child with more than a decent future. If, on the other hand, prospective parents wish to give their child more than a decent future, nothing in Sparrow's argument suggests that this would be objectionable.

Buchanan et al. point out that giving parents the freedom to provide their child with an open future through the use of ART is comparable with the considerable freedom afforded to parents to use a variety of environmental resources 'to attempt

---

[120] Feinberg J (1980) The Child's Right to an Open Future. In: Aiken W, LaFollette H (eds), *Whose Child? Children's Rights, Parental Authority and State Power* (Littlefield, Adams & Co) 126.

[121] Sparrow R (2012) A Child's Right to a Decent Future? Regulating Human Genetic Enhancement in Multicultural Societies. *Asian Bioethics Review* 4: 355, 366.

[122] Ibid.

[123] Ibid.

[124] Ibid.

[125] Ibid.

[126] Feinberg J 'The Child's Right to an Open Future' in William Aiken and Hugh LaFollette (eds), *Whose Child? Children's Rights, Parental Authority and State Power* (Littlefield, Adams & Co, 1980) 151.

[127] Sparrow suggests that even embryos that have a disability may have a 'decent future'. Sparrow R (2012) A Child's Right to a Decent Future? Regulating Human Genetic Enhancement in Multicultural Societies. Asian Bioethics Review 4: 355, 366.

to give their child what they consider to be the best life possible'.[128] However, they also point out that whether the state should protect a right to an open future is controversial, and[129] suggest that in some instances an open future may actually be detrimental to a child's welfare. This is because substantial choice can lead 'to anxiety and indecision', whereas adhering to one path from an early age and following a rigorous training schedule may be more conducive to adult success.[130] Conversely, it may be argued that dictating a child's path from an early age deprives that child of their right to self-determination in much the same way as the Amish children in *Yoder*.[131]

Dena Davis has argued that rather than giving a child an open future, permitting parents to select for abilities may have the opposite effect.[132] If parents have knowledge of their child's strengths even before that child is born, they may only allow the child to undertake activities in which there is a good chance they will excel. This would effectively deny the child valuable experiences that may help shape them into a well-rounded adult. However, this could also be said of parents who realise their child's skills and talents at an early age and is therefore not a good reason to prevent those parents who genuinely wish to give their child an open future the opportunity to do so. Further, if selection is limited to general-purpose means this problem does not arise, as general-purpose means do not lead a child down a specific life path, but would be valuable to a child regardless of the kind of life they choose to lead.[133] Davis argues that the risk of a child's open future being stifled correlates with the prospective parents' investment in their future child's traits.[134] Thus she suggests that the impact on the future child's open future will be different in a situation where prospective parents are using, for example, PGT-M, and polygenic screening is a secondary consideration, compared to a case where PGT-P is used solely for the purposes of selecting a specific trait.[135] In the second scenario, the parental investment, not just in financial terms but also in time, inconvenience and physical discomfort, may cause the parents to feel entitled to a specific result.[136] This is where fertility counselling becomes of crucial importance. Prospective parents need to receive counselling about the importance of keeping their child's future open. They must also be informed that even though an embryo may have a high polygenic score for a particular trait, there

---

[128] Buchanan A et al. (2000) From Chance to Choice: Genetics and Justice. Cambridge University Press, 171.

[129] Ibid.

[130] Ibid.

[131] *Wisconsin v Yoder* 406 US 205 (1972). For a brief summary of this case, please refer to the previous chapter.

[132] Davis D (2009) The Parental Investment Factor and the Child's Right to an Open Future. Hastings Center Report 39(2): 24–27.

[133] I acknowledge that there are no guarantees that even if selection is limited to general-purpose means that parents will not direct their child's pursuits in a particular direction.

[134] Davis D (2009) The Parental Investment Factor and the Child's Right to an Open Future. Hastings Center Report 39(2): 24–27, 25.

[135] Ibid 26.

[136] Ibid.

is no guarantee that their future child will have the phenotype associated with that trait.

Having briefly re-examined Joel Feinberg's open future argument, I suggest prospective parents should have the freedom to choose general-purpose means in their future offspring. By providing their children with an open future, prospective parents are contributing to their future child's right to self-determination. If a range of open futures are available, parents have the freedom to choose which embryo they wish to implant.[137] This decision would be much the same as the decision to send one's child to piano as opposed to violin lessons; both are worthwhile but different options.

6.3.4 The Therapeutic/Non-Therapeutic Distinction

The therapeutic/non-therapeutic distinction, also known as the therapy–enhancement or treatment–enhancement distinction,[138] is the idea that medical procedures, such as PGT, are acceptable for therapeutic purposes, such as preventing disabling conditions, but unacceptable for non-therapeutic purposes, such as selecting the most intelligent child.[139] As the distinction stems from the concept of medical necessity, to qualify as medically necessary a procedure needs to treat or prevent a medical or disabling condition. This distinction can sometimes be difficult to draw, as illustrated by Norman Daniels' growth hormone treatment cases. Daniels introduces his readers to Johnny, 'a short 11-year-old boy with documented [growth hormone] deficiency resulting from a brain tumour. His parents are of average height. His predicted adult height without [growth hormone] treatment is approximately 160 cm'.[140] We are also introduced to Billy, 'a short 11-year-old boy with normal [growth hormone] secretion'.[141] Billy's parents are extremely short and his predicted adult height is the same as that of Johnny. Daniels states that in the US, Johnny's treatment would be covered by his parents' health insurance, whereas Billy's would not. As Daniels notes, both children will be equally disadvantaged by their height and there 'is no reason to think the difference in the underlying causes of their shortness will lead

[137] As not every embryo that is implanted will become a child, more than one embryo may have to be chosen.

[138] The terms 'treatment-enhancement' and 'therapy-enhancement' often appear in American literature. However, they are confusing in the context of selection as embryos are being neither treated nor enhanced.

[139] The distinction between therapeutic and non-therapeutic treatment is also used in other areas of law and medicine, see, eg, *Marion's Case* (1992) 175 CLR 218.

[140] Daniels N (2000) Normal Functioning and the Treatment-Enhancement Distinction. Cambridge Quarterly of Healthcare Ethics 9: 309, 311.

[141] Ibid.

people to treat them in ways that make[s] one happier or more advantaged than the other'.[142] [143]

What would be the choices available to Johnny and Billy's parents if they underwent IVF with PGT? If a genetic basis for brain tumours in children were to be discovered, Johnny's parents would have the freedom to test for and select against this pre-disposition. As a consequence, the embryo that became Johnny would be of average height. Billy's parents, on the other hand, would not be permitted to test their embryos for genes related to height, because being of below average height is not a genetic or disabling condition.[144] These cases illustrate the arbitrariness of the therapeutic/non-therapeutic distinction. Avoiding the birth of a child with a genetic pre-disposition to cancer, or treating the side effects of cancer, fall within the sphere of what is medically necessary. Being short, on the other hand, is not considered to be a disabling condition and parents seeking to give their child the advantage of average or above average height are seen as trying to confer a benefit on their child. Is there really a moral difference between these two cases? If so, does this difference represent a valid reason to prevent parents selecting for desirable characteristics?

The therapeutic/non-therapeutic distinction is seen by some to represent the moral boundary between procedures that are permissible and those that are impermissible.[145] David Resnik notes that one reason therapeutic procedures are considered moral is because they promote 'the goals of medicine'.[146] Conversely, non-therapeutic procedures may be perceived as having 'morally questionable goals'.[147] However, supporters of the distinction are willing to concede that this boundary is not set in stone.[148] Resnik asserts that the most that can be said in support 'of the distinction is that it defines moral zones without any sharp boundaries'.[149] Further, he suggests that rather than asking whether a procedure is therapeutic or non-therapeutic, consideration should be given to the risks[150] posed by the procedure and the benefits offered.[151]

Applying these considerations to embryo testing, it can be said that the risks that would be involved in selecting for abilities would be no different to those that may transpire in the course of selecting against disabling conditions. As I have already concluded that using PGT for negative selection is acceptable, there seems no reason

[142] Ibid.

[143] Daniels N (2000) Normal Functioning and the Treatment-Enhancement Distinction. Cambridge Quarterly of Healthcare Ethics 9: 309, 309.

[144] One exception would be cases where the embryo will become a person with achondroplasia.

[145] Resnik DB (2000) 'The Moral Significance of the Therapy-Enhancement Distinction in Human Genetics' 9 *Cambridge Quarterly of Healthcare Ethics* 365.

[146] Ibid 368.

[147] Ibid.

[148] Ibid 365.

[149] Ibid 374.

[150] In addition to physical risks, Resnik also considers risks to society, such as those posed by eugenics (ibid 373). Eugenics was considered in the previous chapter.

[151] Ibid 374.

to prohibit the use of embryo testing to select for abilities on the grounds that it is too risky. Further, the benefits of using embryo testing in this way were discussed earlier in this chapter, for example in the context of screening embryos for IQ. The advantages presented by higher levels of intelligence are beneficial whether or not a competitive advantage is conferred (e.g., as intelligence positively affects health outcomes,[152] it is better for both individuals and society as a whole if everyone was healthier).

In a similar vein to Resnik, Buchanan and colleagues argue that 'No reasonable defense of the treatment/enhancement distinction is possible if we expect too much of it'.[153] Further, they note that although the distinction may not offer a straightforward test for determining which uses of PGT should be permitted, there are compelling reasons to believe that some non-therapeutic uses of PGT may give rise to 'serious problems not posed by treatments'.[154] One example cited by Buchanan and his colleagues is the issue of competitive advantage, mentioned earlier in this chapter. They note that if everyone selects in favour of a particular ability, the advantage conferred by that ability ceases to exist. However, as noted earlier in this chapter, this is not a persuasive concern in the context of embryo testing because parents would be testing and selecting for genetic characteristics already present in the embryo.

In contrast to Resnik's modest defence of the therapeutic/non-therapeutic distinction, Isabel Karpin and Roxanne Mykitiuk argue that the 'distinction is inadequate and unhelpful to guide ethical analysis and medical and regulatory decision making'.[155] They suggest that one example of where the distinction struggles is in the context of carrier embryos[156] and question whether selecting against carrier embryos is really therapeutic when the purpose of such selection is to ensure the future child does not themselves face a reproductive dilemma.[157] As selection against carrier embryos is acceptable in many regulatory frameworks, this illustrates that non-therapeutic uses of PGT are already taking place.[158] Further, Karpin and Mykitiuk note that the therapeutic/non-therapeutic distinction is effectively part of the law in jurisdictions which adopt a threshold of seriousness.[159] As I argued in the

[152] Plomin R, von Stum S "The New Genetics of Intelligence," Nature Reviews: Genetics 19, no 3 (2018): 148–159, 148. https://doi.org/10.1038/nrg.2017.104.

[153] Buchanan A et al. (2000) From Chance to Choice: Genetics and Justice. Cambridge University Press, 119.

[154] Ibid 154.

[155] Karpin I, Mykitiuk R (2008) Going out on a Limb: Prosthetics, Normalcy and Disputing the Therapy/Enhancement Distinction. Medical Law Review 16: 413, 413.

[156] Carrier embryos are embryos with the genotype for an autosomal recessive condition (e.g., cystic fibrosis); but once the embryo becomes a child, that child will themselves not have the illness. However, they may pass the disease onto their offspring.

[157] Karpin I, Mykitiuk R (2008) Going out on a Limb: Prosthetics, Normalcy and Disputing the Therapy/Enhancement Distinction. Medical Law Review 16: 413, 434.

[158] For example, this use of PGT is not prohibited by the NHMRC Ethical Guidelines or by any of the statutes regulating ART in Australian states.

[159] Karpin I, Mykitiuk R (2008) Going out on a Limb: Prosthetics, Normalcy and Disputing the Therapy/Enhancement Distinction. Medical Law Review 16: 413, 436.

previous chapter that there is no good reason to limit the use of PGT-M to serious conditions, this may be the first step in blurring the divide between therapeutic and non-therapeutic uses of PGT-P.

The therapeutic/non-therapeutic distinction does not represent a sound moral reason to prevent prospective parents from using embryo testing to select for abilities. While some academics have argued that the distinction may be of some use, its inherent arbitrariness prevents it from being a useful tool in the regulation of embryo testing.

6.3.5 *The Expressivist Objection*

In Chap. 5, the expressivist objection to selection against disability was introduced. This is the argument that using PGT to test for and select against disability reflects negative or discriminatory attitudes towards people with disabling conditions.[160] In the context of selection for polygenic traits, the expressivist objection may suggest that people who are lacking the abilities for which PGT-P is being employed, or who possess them to a lesser extent, are less valuable, or at the very least less fortunate. It may also suggest that the value of the future child is dependent on them possessing a particular trait, for example a high IQ, and in the absence of this trait they may not have been wanted.[161]

Dan Brock argues that it is not specifically PGT that sends these hurtful messages to individuals, but rather the 'underlying attitude or value judgment that these are desirable traits to have' and those who do not have them are in some way lacking .[162] Further, Brock notes that the attitude that it is desirable to have certain abilities, such as those referred to earlier in this chapter as general purpose means, is justified and on the whole uncontroversial.[163] Consequently, as it is often necessary to make such evaluations, it is inevitable that those who are perceived to be less able may hear something upsetting.[164] As Brock points out,the only way to avoid sending a potentially distressing message is to refrain entirely from making any judgments regarding which traits are desirable.[165] As it is impossible and undesirable to avoid making any judgments of this kind,[166] the expressivist objection is not a valid argument to prevent prospective parents from using PGT-P to select for general-purpose means in their future offspring.

---

160 See, eg, Parens E, Asch A (1999) The Disability Rights Critique of Prenatal Genetic Testing. 29(5) *Hastings Center Report* 29(5): S1, S2.

161 Parens and Asch have made this point in the context of prenatal testing for disability. Ibid.

162 Brock DW (2009) Is Selection of Children Wrong? in Savulescu J, Bostrom N (eds) *Human Enhancement* (Oxford University Press) 251, 264–265.

163 Ibid 265.

164 Ibid.

165 Ibid.

166 Ibid.

6.3.6 Playing God

Coady describes the playing God argument as the idea 'that there are certain things that it is presumptuous for human beings to undertake because those matters are really in the care of God'.[167] One such matter is the genetic makeup of our offspring. To intervene in this sphere would therefore constitute hubris. Coady explains that in Christianity, God is considered all-powerful, all-knowing and supremely benevolent.[168] Humans on the other hand, are fallible and only possess limited power and benevolence. Thus Coady suggests that the playing God argument 'can be interpreted as going beyond the limits we have' and 'acting in ways that ignore the in-built constraints on our knowledge, power and benevolence'.[169] This section asks, is the playing God argument a satisfactory ground for curtailing reproductive liberty?

Dan Brock points out that if the playing God argument is an objection to interfering with natural processes under God's sole control,

> then it is of course notoriously difficult to see to which human interventions this objection should apply. Building airplanes and cars enable humans to travel long distances that they could not manage without these interventions, yet the playing God objection is rarely applied to them.[170]

If the playing God argument alone is persuasive enough to prevent the use of PGT-P, should it also be applied to stop companies building aircraft, cars and even computers? Brock notes that as God has given humans the ability to create a range of technological advances, it is unclear why it would be contrary to God's will to make use of them'.[171] Therefore, Brock suggests that if God gives humankind the ability to test for abilities in embryos, it is not clear why using this knowledge would run contrary to God's will.[172] Furthermore, Brock asserts that intervening 'in natural processes for the betterment and benefit of humans is a fundamental human capacity and propensity, and there seems to be no principled line to demarcate when doing so would be usurping God's prerogatives'.[173]

Similarly, Peters argues that human beings are 'created co-creators'.[174] Peters suggests that this term is significant in two important respects. Firstly, it reminds us that God created us and therefore any creativity we possess cannot be considered on a par with that of God.[175] Secondly, the reference to co-creator signifies that human

---

[167] Coady C A J (2009) Playing God. In: Savulescu J, Bostrom N (eds), *Human Enhancement* (Oxford University Press) 155, 155.

[168] Ibid 163.

[169] Ibid.

[170] Brock DW (2009) Is Selection of Children Wrong? In: Savulescu J, Bostrom N (eds) *Human Enhancement* (Oxford University Press) 251, 267.

[171] Ibid.

[172] Ibid.

[173] Ibid.

[174] Peters T, '"Playing God" and Germline Intervention' (1995) 20 *Journal of Medicine and Philosophy* 365, 377.

[175] Ibid.

beings have a role to play in continuing God's work and may be able to exert some influence over the direction creation takes.[176] As Coady notes, viewed in this context, 'Playing God is no accusation'.[177]

A further reason the playing God argument is not an appropriate justification for preventing parents testing embryos for positive characteristics, is that it seeks to impose on society specific religious views that not everyone shares. For those who do not believe in God, the charge of playing God is meaningless.

6.4 Conclusion

This chapter has argued that prospective parents should be able to exercise their reproductive liberty through the use PGT-P to select against polygenic disabling conditions. While this technology is still in its infancy and there are understandably concerns about its use, much of this apprehension is related to the novelty of PGT-P. For example, concerns about the under-representation of people with non-European ancestry in biobanks and the negative impact this has on the accuracy of PGT-P for such populations will cease to be problematic as more data is collected and more research is undertaken.[178] Therefore, rather than this being a moral objection to the use of PGT-P, it is an objection to its premature use. In a similar vein, the novelty of the technology has led to concerns about whether the results of PGT-P will be too complicated for patients to understand. While the options presented by PGT-P may indeed increase the complexity of the decision-making process for prospective parents, it is possible to identify areas of concern and develop comprehensive genetic counselling programs to address these.[179] As PGT-P will not only be new to prospective parents, but also genetic counsellors and other healthcare providers, there is also an opportunity to create training programs focused on PGT-P. Similarly, the anxieties that exist around the use of polygenic risk scores to predict genetic conditions in embryos is yet another area in which research can be undertaken. As PGT-P is already used overseas, research studies following live births will be able to demonstrate its accuracy, or lack thereof. Therefore, concerns around the use of PGT-P to select against polygenic conditions may be ameliorated in a number of ways, including public education campaigns, regulation and thorough training programs for staff involved in the delivery of PGT.

---

[176] Ibid.

[177] Coady CAJ (2009) Playing God. In: Savulescu J, Bostrom N (eds), *Human Enhancement* (Oxford University Press) 155, 156.

[178] See, eg, Lello L, Tellier LCAM (2021) Polygenic Risk Scores, PGT-P, and IVF: Sibling and Non-European Ancestry Validations. Fertility and Sterility 116(3): e20.

[179] Turley et al. have outlined a number of areas where confusion may arise in the context of PGT-P, together with recommendations for responsible communication. Turley P, Meyer M, Wang N et al. (2021) Problems with Using Polygenic Scores to Select Embryos. N Engl J Med 385(1): 78–86, 85.

Further, should the technology become available in Australia, prospective parents should have the freedom to test and select for abilities in their future offspring. A good starting point for regulation in this area would be to limit testing to what have been referred to as general-purpose means. This would provide prospective parents with the freedom to test and select for traits that would prove valuable to a child, regardless of what kind of life that child chose to lead. By permitting selection for general-purpose means, as opposed to career specific or lifestyle specific characteristics, a child's right to an open future is being respected. As protecting a child's right to an open future protects their right to self-determination, and as no conclusive evidence exists to show that having an open future is negative for a child, parents should have the freedom to select in favour of general-purpose means, so as to provide their child with an open future.

Permitting selection for general-purpose means also gives parents the opportunity to act in a procreatively beneficent manner and thus act morally by contributing to the welfare of their future child. While it may not be possible for prospective parents to choose the 'best life' for their future child, as ranking lives is inherently problematic, they may be able to select the embryo which is most likely to develop into a child with a range of general-purpose means. This chapter has also examined the therapeutic/non-therapeutic distinction, the expressivist argument, and the playing God argument, to see if they provide justifications for restricting prospective parents' reproductive liberty. I concluded that none of these are useful regulatory tools.

This chapter has shown that prospective parents should be permitted to use PGT-P to select against polygenic conditions. It has also highlighted that there are good reasons for children to begin their lives with characteristics that will help them to live a good life and consequently, good reasons for parents to have the freedom to use PGT-P in order to select for such characteristics. Therefore, to allow prospective parents to exercise their reproductive liberty in these ways, the regulatory framework needs to be responsive to new applications of embryo testing, such as PGT-P.

Chapter 7
The Difficult Cases: Selecting for Disabling Conditions, Non-medical Sex Selection and Tissue Typing

Abstract This chapter explores three uses of embryo testing, which I have termed the 'difficult cases'. The difficulty arises because the choices the prospective parents wish to make do not, at least at first glance, align with the welfare interests of the child who will be born. The uses of embryo testing under consideration are selection for disabling conditions, non-medical sex selection, and human leukocyte antigen (HLA) typing for the creation of 'saviour relatives'. The chapter begins by reviewing the concept of reproductive liberty and introducing Derek Parfit's non-identity problem. I argue that while reproductive liberty is expansive, it must be balanced against the welfare interests of the future child.

7.1 Introduction

This chapter considers the potential conflict between reproductive liberty and the welfare of the future child. As discussed in chap. 4, when prospective parents make decisions about the kind of child they wish to have through assisted reproductive technology (ART), they are exercising their reproductive liberty. Embryo testing may give rise to a conflict between the reproductive liberty of prospective parents and the welfare of the future child in situations where parents wish to test for characteristics that may be perceived as not wholly conducive to the welfare interests of the child to be born. Therefore, consideration needs to be given to how these values are to be reconciled in the development of a revised framework for embryo testing. The chapter starts by briefly reviewing the concept of reproductive liberty which was discussed in detail in chap. 4. It then examines what is meant by the term 'welfare of the child' and why it is important in the regulation of embryo testing. The chapter then illustrates the conflict between reproductive liberty and the welfare of the child by reference to three uses of embryo testing that were not discussed in the previous two chapters. The first case that will be considered is selection for disabling conditions.[1]

[1] As noted in chap. 4, the term 'disabling condition' is used in preference to 'disability' to ensure that conditions which are often thought of as illnesses (e.g. cystic fibrosis) are brought within the concept of disability used in this book.

M. de Souza, *The Regulation of Embryo Testing in Australia*,
https://doi.org/10.1007/978-981-96-5902-9_7

The second, sex selection for non-medical reasons, and lastly, I consider this conflict in the context of human leukocyte antigen typing (HLA typing). The chapter will then conclude with suggestions for law reform in this area.

7.2 Reproductive Liberty and Difficult Cases

The theory of reproductive liberty used in this book is based on John Robertson's concept of procreative liberty.[2] Reproductive liberty includes the freedom to undertake activities or make decisions that culminate in the birth of a child. This freedom may only be limited in cases where the exercise of this liberty will cause actual harm to another person.[3] As discussed in chap. 4, Robertson's theory has been chosen because it conceptualises procreative liberty as a negative claim right with a broad scope. In this way it opens the door to discussions about a range of uses of PGT.[4] Further, given the continuous advancements in PGT, it is desirable to utilise a model that demands more than merely 'theoretical or speculative' harm to limit reproductive liberty. In this section, I explore whether reproductive liberty extends to selection for disabling conditions, HLA typing, and non-medical sex selection.

A particular use of ART is an exercise of reproductive liberty if it is 'essential or material' to the values that underpin the prospective parents' decision to have a child.[5] This is an illustration of the expansive nature of Robertson's conception of procreative liberty. Further, on Robertson's account of procreative liberty, a 'certain degree of conformity with common understandings of why reproduction is important' is required.[6] On this basis, an argument can be advanced for including selection for disabling conditions within the scope of reproductive liberty. Parents who wish to select an embryo with a disabling condition often do so with the intention of selecting characteristics that would allow a child to fit into the parents' lifestyle and community. Therefore, selection for disabling conditions falls within the scope of reproductive liberty because the freedom to bring up a child in line with one's own values may be considered a central aspect of parenthood. It should, however, be borne in mind that falling within the scope of reproductive liberty does not mean a practice should be permitted, simply that there need to be good reasons to prevent it.[7] This will be examined further later in this chapter.

---

[2] See chap. 4 [4.3.1].

[3] For a detailed discussion see chap. 4 [4.3].

[4] See chap. 4 [4.3.3].

[5] Robertson JA (2003) 'Procreative Liberty in the Era of Genomics' American Journal of Law and Medicine 29: 439, 449.

[6] Ibid.

[7] Not all theories of reproductive liberty lead to the same conclusion. E.g. if one were to adopt Dillard's theory of reproductive liberty (see chap. 4 [4.3.2.2]), which proffers a far more limited version of reproductive rights, selection in favour of disabling conditions would fall outside its scope. This is because selecting a child with a particular disability, regardless of the parents' reasons for their choice, is not fundamental to the continuation of the species.

As explained in chap. 1, HLA typing, also known as tissue typing, gives the parents of a sick child the opportunity to have a second child whose tissue type matches that of the existing child.[8] Once the child is born, their umbilical cord stem cells are donated in the hope that the transplant will cure the sick child. HLA typing may fall within the scope of reproductive liberty because the possibility of creating a 'saviour sibling' for an existing sick child could play a determinative role in a couple's decision to have another child. Even though a couple may have intended to have another child, being denied the use of HLA typing may cause them to re-consider their decision and focus their attention on finding a cure for their existing ill child. In this way, reproductive liberty plays a decisive role in their reproductive decision-making.[9]

Having concluded that HLA typing and using PGT-M to positively select for disabling conditions both fall within the scope of reproductive liberty, I turn to non-medical sex selection. Non-medical sex selection does not fit easily within the scope of reproductive liberty. This is because most people would not consider the freedom to have a child of a particular sex to be essential to their decision to procreate. However, as John Harris notes, no general consensus exists regarding the extent or nature of reproductive liberty and the central theme of theories in this sphere is 'respect for autonomy and for the values which underlie the importance attached to procreation'.[10] Further, he questions whether prospective parents' reproductive choices should simply be cast aside 'as a simple and idle exercise of preference'.[11] Consequently, it is perhaps presumptuous to consider non-medical sex selection as falling outside the scope of reproductive liberty simply because the majority of prospective parents may not consider the sex of their child to be of vital importance.

7.3 *The Welfare of the Future Child*

This section considers the welfare of the future child and the role this concept should play in decision-making with regard to specific uses of PGT. Consideration of the future child's welfare may justify curtailment of prospective parents' reproductive liberty. If the welfare of a future child will be stifled due to the circumstances surrounding their conception, this may be a good reason to prevent parents having

---

[8] For a detailed discussion of HLA typing, see chap. 1 [1.3]. The National Health and Medical Research Council's Ethical Guidelines on the Use of Assisted Reproductive Technology in Clinical Practice and Research 2017 also permit the creation of a 'saviour relative'—see [8.17] of the guidelines.

[9] Not everyone agrees that this should be the case. E.g. Maura Ryan argues that by including HLA typing within the scope of procreative liberty, Robertson treats future children as property or objects that can be acquired by another person, and thus 'fails to respect offspring as autonomous beings, as ends in themselves': Ryan MA (1992) The Argument for Unlimited Procreative Liberty: A Feminist Critique. In: Campbell CS (ed) What Price Parenthood? Ethics and Assisted Reproduction. Dartmouth, 84, 85.

[10] Harris J (2005) Sex Selection and Regulated Hatred. Journal of Medical Ethics 31: 291, 293.

[11] Ibid.

access to certain uses of embryo testing. However, some academics argue that as long as a child has a life worth living they are not harmed by being born.[12] The reasoning behind this is that the alternative to having a life of less than optimum wellbeing is non-existence, and there is a presumption that being born is preferable to not existing at all. This is also the attitude that has been taken in other contexts, such as wrongful life cases. In *Harriton v Stephens*,[13] the High Court considered an action for damages brought by a young woman with significant disabilities. Alexia Harriton suffered from severe mental retardation, blindness and deafness; congenital disabilities caused by the rubella virus. Harriton argued that had her mother been correctly diagnosed with rubella and informed that this would cause significant abnormalities in the foetus, she would have terminated her pregnancy. As a result, Alexia would not have been born and would therefore not have severe disabilities.

By a majority of 6–1, the High Court in *Harriton* held that the appellant had not suffered actionable damage, because a life with severe disabilities could not be compared to non-existence. Describing the comparison as 'impossible', Crennan J noted that Harriton's disabilities could not be described as 'a loss, deprivation or detriment' if compared with the alternative, which would have been non-existence.[14] Thus, the majority were unwilling to consider life itself as damage, suggesting that existence is preferable to non-existence.

In the context of PGT, there is, however, another perspective: Derek Parfit's 'Same Number Quality Claim'.[15] We can rely on this claim because in discussions about embryo testing, questions tend to be centred on what sort of children there should be, as opposed to whether or not there should be a child. Parfit suggests that if in a given situation there are two possible outcomes and regardless of which outcome transpires the same number of people would live, 'it would be worse if those who live are worse off, or have a lower quality of life, than those who would have lived'.[16] Parfit notes that the Same Number Quality Claim does not imply that a person's life is 'intrinsically morally undesirable', simply that had an alternative child been born they would have had a better life.[17] Applied to PGT, the Same Number Quality Claim can act as a guide to whether tests such as PGT-M to select for disabling conditions, are permissible.

References to the welfare of the child can be found in the statutes of all six of the states which have enacted ART legislation, as well as in the National Health and Medical Research Council's Guidelines. The *Assisted Reproductive Treatment Act 2008* (Vic) s 5(a) provides that 'the welfare and interests of persons born or to be born as a result of treatment procedures are paramount'. Similarly, the *Assisted*

[12] See, eg, Cohen IG (2011) Regulating Reproduction: The Problem with Best Interests. Minnesota Law Review 96: 423, 437; Harris J (2000) The Welfare of the Child. Health Care Analysis 8: 27, 29.

[13] *Harriton v Stephens* (2006) 226 CLR 52.

[14] Ibid [252]–[253].

[15] Parfit D (1986) Reasons and Persons. Oxford University Press, 360.

[16] Ibid.

[17] Ibid.

Reproductive Treatment Act 1988 (SA) s 4A states that the 'welfare of any child to be born … must be treated as being of paramount importance'. The *Human Reproductive Technology Act 1991* (WA) does not necessitate that the future child's welfare be the paramount consideration, but requires that 'the prospective welfare of any child to be born … is properly taken into consideration'.[18] The *Assisted Reproductive Technology Act 2007* (NSW) s 3(b)(i) uses different terminology, but to the same effect, as it seeks to 'protect the interests of… a person to be born as a result of ART treatment'. In a similar vein, the *Assisted Reproductive Technology Act 2024* (ACT) s 8(a) also refers to the protection of the interests of the future child.[19] Last, the National Health and Medical Research Council's *Ethical Guidelines on the use of Assisted Reproductive Technology in Clinical Practice and Research 2017* provide that the 'interests and wellbeing of the person who may be born … must be an important consideration' in all treatment decisions.[20]

The fact that all Australian ART statutes, as well as the NHMRC Guidelines, require the future child's welfare to be taken into account, indicates the importance of this concept in the regulatory framework for ART. In the following sections, three uses of embryo testing will be considered and their impact on the welfare of the future child discussed. Only one of these three uses (i.e. HLA typing) is currently permitted in Australia.

7.4 Selecting for Disabling Conditions

This section considers whether prospective parents should be permitted to use PGT to select for disabling conditions in their offspring. At present, the Australian regulatory framework prohibits such uses of PGT, with the NHMRC Guidelines providing that PGT 'may not be used to preferentially select in favour of a genetic condition, disease or abnormality' in the person to be born.[21] Is this absolute prohibition necessary, or would a more nuanced approach be more appropriate? To attempt to answer this question, I will first consider the conflict between reproductive liberty and the welfare of the child in this context. I will argue that parents should not be permitted to select for disabling conditions when 'healthy' embryos are available.

It is useful at the outset to set clear parameters for this discussion. It will be argued that in cases where parents have a selection of embryos to choose from, they should not be permitted to select an embryo that will become a child with a disabling condition. However, in cases where only one embryo is created, or where all the embryos have a disabling condition, prospective parents may choose to implant an embryo that will become a child with a disability. This is because prohibiting parental

[18] *Human Reproductive Technology Act 1991* (WA) s 4(1)(d)(iv).

[19] Similar terminology is used in the *Assisted Reproductive Technology Act 2024* (Qld) s 3(1)(a)(i).

[20] National Health and Medical Research Council, *Ethical Guidelines on the use of Assisted Reproductive Technology in Clinical Practice and Research 2017* ('*NHMRC Guidelines*') 20.

[21] Ibid [8.15.2].

choice in this situation would be akin to requiring parents to terminate a pregnancy where pre-natal diagnosis has shown that the baby will be born with a disability. As many would find this conclusion repugnant, parents must be permitted to implant an embryo with a disabling condition where no other embryos exist. In cases where multiple embryos have been created and all embryos have the genetic makeup to become a child with a disabling condition, parental choice should be restricted to selection of the embryo with the least severe condition.

The issue of selection for disability gained prominence in 2002, after a deaf lesbian couple in the United States used a sperm donor with a family history of deafness to deliberately create a deaf child. The couple (Sharon Duchesneau and Candy McCullough) already had a deaf daughter and wanted their second child to also be part of their deaf culture. The couple, like many in the deaf community, did not consider deafness as a disability; instead, viewing it as a cultural identity.[22] However, deafness is only one possible disabling condition for which parents may choose to select. As Julian Savulescu points out, couples with achondroplasia (a genetic disorder that results in dwarfism) may wish to have a child with the condition and prospective parents with an intellectual disability 'may wish to have a child like them'.[23]

In chap. 4, various models of disability were considered and I suggested that the interactional model of disability was the most suitable to use when thinking about reproductive choices. As discussed in that chapter, thinking about disabling conditions as interactions between impairment and environmental factors gives the concept a fluidity that is vital for making decisions in this context.

To justify a prohibition on selection for disabling conditions, such as that currently imposed by the NHMRC Guidelines, it needs to be asked whether being born with a disability harms the future child. If it can be argued that it does, this satisfies John Stuart Mill's Harm Principle and thus justifies a state-imposed prohibition. As discussed in chap. 4, interference with the reproductive liberty of prospective parents is only legitimate if it prevents harm to others.

7.4.1 Does Selection for Disabling Conditions Harm the Future Child?

Whether a particular use of PGT can harm a future child requires consideration of two matters. First, what is meant by the term 'harm'? Second, can a future child be harmed?

---

[22] Spriggs M (2002) Lesbian Couple Create a Child who is Deaf Like Them. Journal of Medical Ethics 28: 283, 283.

[23] Savulescu J (2002) Deaf Lesbians, "Designer Disability," and the Future of Medicine. British Medical Journal 325: 771, 771.

As Joel Feinberg notes, the 'word "harm" is both vague and ambiguous'.[24] While there are many possible definitions of harm, in this book harm will be defined as a setback to interests;[25] specifically, the interests of the future child. Whether a future child's interests have been harmed by their parents' decision to use PGT to select for a disabling condition depends on whether 'that interest is in a worse condition than it would otherwise have been in' had PGT not been used.[26] Feinberg suggests that a person's 'interests' include their physical health and fitness, 'the integrity and normal functioning of [their] body', the absence of severe pain or suffering, as well as cognitive and mental capabilities conducive to the pursuit of that person's goals.[27] Consequently, Feinberg terms these interests 'welfare interests'.[28] Feinberg explains that the importance of welfare interests lies in their role in allowing a person to reach their ultimate goals, whether that goal is to write a bestselling novel, create a work of art, or secure a well-paid job.[29] Therefore, any interference with a person's welfare interests 'is the most serious … kind of harm a person can sustain'.[30]

7.4.1.1 Can a Future Child Be Harmed?

In his non-identity problem,[31] Derek Parfit argues that our identity depends on when we were conceived and therefore if we were conceived at a different time, from different gametes, we would be a different person. In the context of PGT, this means that our identity is based on the embryo from which we developed and if a different embryo were selected, we would not have been born. Consequently, being born with a disabling condition does not harm the future child, for if a different embryo is selected, a different child would be born. Accordingly, if having a disability does not harm the future child, unless of course the disabling condition is so bad that it would be better if they had never been born, why should parents be prohibited from selecting for disabling conditions?

Consider the following situation: prospective parents create two embryos, and testing reveals one will become a child with a disabling condition and the other will not. The couple decide to implant the embryo which will become the child with the disabling condition. Parfit suggests that the future child cannot complain that they have been harmed because if the 'healthy' embryo had been implanted in its place, that child would not exist. Thus the prospective parents' decision is non-person

[24] Feinberg J (1987) The Moral Limits of the Criminal Law Volume 1: Harm to Others. Oxford University Press, p 32.

[25] Ibid p 34.

[26] Ibid p 34.

[27] Ibid p 37.

[28] Ibid.

[29] Ibid.

[30] Ibid pp 37–38.

[31] Parfit D (1986) Reasons and Persons. Oxford University Press, 359.

affecting.[32] To explain why non-person affecting principles can be used to justify prohibiting selection for disability, I turn to the notion of 'victimless harm'.[33]

The idea of victimless harm gives rise to the argument that in cases where two possible outcomes result in the same number of individuals being brought into existence, the morally preferable outcome is the one in which those individuals enjoy a higher quality of life.[34] This principle appears to provide the basis for defending a prohibition on selection for disabling conditions. Parfit refers to this principle as 'The Same Number Quality Claim'.[35] Based on this principle, prospective parents should not be permitted to select embryos that will develop into a child with a genetic condition, when an embryo that will develop into a child without that condition is available. Philip G Peters notes that 'conduct can be harmful even though it makes no specific child worse off than he or she would otherwise have been. It is harmful because it causes unnecessary suffering'.[36] Therefore, by selecting an embryo that will become a child with a disabling condition, a child is unnecessarily harmed and this harm can be avoided by selection of an embryo without that condition. Peters asserts that given the power of this claim 'it is remarkable that so obvious a proposition has not already become part of the fabric of reproductive decision-making'.[37] Further, he remarks that it is surprising that Parfit's claim has never been 'explicitly relied upon... to defend the regulation of reproductive technology'.[38]

In a posthumously published paper,[39] Parfit revisits the non-identity problem, including what he refers to as Pure Same Number Cases. These are the kind of scenarios that are encountered in the sphere of embryo testing.[40] Parfit states that 'it would be in itself better if the people who existed were the ones whose lives would be more worth living'.[41] Parfit accepts that to solve the non-identity problem, it needs to be explained why certain acts (for example, selecting an embryo that will become a child with a disabling condition) are wrong, even though no actual people are worse

---

[32] An example of a person affecting case is if a woman is pregnant with a child and she has the option to take a pill that will cure that child of a genetic disorder. In this situation, the same person will be born whether or not the pill is taken, the decision affects whether or not that person has a genetic disorder.

[33] Peters PG (1998) Harming Future Persons: Obligations to the Children of Reproductive Technology. Southern California Interdisciplinary Law Journal 8: 375, 395.

[34] Parfit D (1986) Reasons and Persons. Oxford University Press, 360.

[35] Ibid. This principle was introduced in [7.3].

[36] Peters PG (1998) Harming Future Persons: Obligations to the Children of Reproductive Technology. Southern California Interdisciplinary Law Journal 8: 375, 395.

[37] Peters PG (2009) Implications of the Non-identity Problem for State Regulation of Reproductive Liberty. In: Roberts MA, Wasserman DT (eds) Harming Future Persons. Springer, 317, 325.

[38] Ibid.

[39] Parfit D (2017) Future People, the Non-identity Problem, and Person-affecting Principles. Philosophy & Public Affairs 45(2): 118.

[40] Cf. mixed same number cases. These are cases where some of the same people may exist in each of two scenarios.

[41] Parfit D (2017) Future People, the Non-identity Problem, and Person-affecting Principles. Philosophy & Public Affairs 45(2): 118, 120.

off.[42] Parfit's theory suggests that it would be wrong to deliberately select an embryo which will become a child with a disabling condition, if 'some other possible act would not be worse for anyone'.[43] To explain why this is the case, Parfit turns to the Wide Principle. He argues that out of two possible outcomes, the less desirable outcome is the one which 'would be less good for people'.[44] By this, he means that the outcome which would provide greater benefit should be preferred. Parfit asserts that even if an outcome would not be worse for anyone, 'this outcome may be less good for people than the other outcome would have been for the different people who, in this other outcome, would have existed'.[45] Applied to PGT, if a child with a disabling condition is created and that child will, to use Parfit's example,[46] endure decades of suffering, that child's life 'would be *bad* for him, and his non-existence would not have been *bad* for him'.[47] This is because a disabling condition is, in itself, negative in nature,[48] whereas non-existence of itself is not.

7.4.1.2 Critiquing the Victimless Harm Justification

Eve Garrard and Stephen Wilkinson argue that defending the prohibition on selection for disabling conditions on the basis of victimless harm results in a curious situation.[49] If it is argued that pursuant to the Same Number Quality Claim embryos expected to have a higher quality of life should be selected over embryos which are expected to have a lower quality of life, this suggests that embryos with desirable traits should be selected over average or 'normal' embryos. While in chap. 6. I argued that parents should have the freedom to make such decisions, the victimless harm justification for prohibiting selecting embryos with a genetic condition necessitates a mandatory requirement that parents should select the embryo that is expected to have the highest level of wellbeing. As a solution, Garrard and Wilkinson suggest an approach based on what they have termed 'prioritarianism'[50] together with the Same Number Quality Claim. A prioritarian approach would advocate helping 'those who are worse off [rather] than those who are better off even when this generates the same improvement

[42] Ibid 126.

[43] Ibid 128.

[44] Ibid 129.

[45] Ibid 130.

[46] Ibid 132.

[47] Ibid 133.

[48] For a discussion of disability theory, see chap. 4 [4.2].

[49] Garrard E, Wilkinson S (2006) Selecting Disability and the Welfare of the Child. The Monist 89: 482, 493–494.

[50] Ibid 496. This term is adapted from Derek Parfit, who refers to this kind of thinking as the 'priority view'. See Parfit D (1997) Equality and Priority. Ratio 10: 202, 213.

in welfare'.[51] As Parfit states, '[w]e should not give equal weight to equal benefits, whoever receives them. Benefits to the worse off should be given more weight'.[52]

Applied to PGT, the prioritarian approach would advocate bringing people into existence 'who will be higher up the welfare scale rather than those who will be lower down. But the strength of these reasons wanes the further up the scale we go'.[53] Using this perspective, prioritarianism supports a prohibition on selecting for disabling conditions while simultaneously distinguishing imposing upon prospective parents a mandatory requirement to select the 'best' embryo. Therefore, adopting the priority view, it can be argued that there are good moral reasons to prevent parents deliberately selecting children with disabling conditions, but far fewer reasons to require parents to select embryos that will have a high level of wellbeing.

Glenn Cohen criticises the prohibition against selecting for disability for having a eugenic basis, because 'non-person affecting' principles are being used to dictate who is brought into this world.[54] While parents may seek to avoid disabling conditions in their future offspring, the freedom to have a child with a disability is not available to parents under the current regulatory framework. Therefore, the choice of which embryos may be selected is being made by the state. There are, however, important ways in which the current regulatory framework differs from the state intervention associated with 'old eugenics' programmes. First and most importantly, the motivation behind the current framework is the prevention of harm and disadvantage to children. This can be illustrated by the provisions contained in both the NHMRC Guidelines and the state legislation discussed in the previous section. In contrast, the motivation behind, for example, the Nazi eugenic programme was to prevent certain kinds of people being born. Secondly, unlike historical eugenics programmes, the current framework would permit implantation of an embryo that will become a child with a disabling condition where no 'healthy' embryos are available. This shows that while in principle no objection exists to implanting embryos that have an impairment, there is a presumption in favour of minimising suffering and limitations for those who will be born.

Trying to ascertain whether selection for a disabling condition harms the future child is complicated, as the above discussion indicates. However, it is also important, because constraints on the reproductive liberty of prospective parents are only permissible if they prevent actual harm (i.e. it sets back the welfare interests of the future child). Having explored various perspectives on this issue, the question remains unanswered. This highlights one of the difficulties with the non-identity problem. By arguing that no one is harmed by the decision to select one embryo over another, the interests of the future child can be set aside, and prospective parents can enjoy

---

[51] Garrard E, Wilkinson S (2006) Selecting Disability and the Welfare of the Child. The Monist 89: 482, 496.

[52] Parfit D (1997) Equality and Priority. Ratio 10: 202, 213.

[53] Garrard E, Wilkinson S (2006) Selecting Disability and the Welfare of the Child. The Monist 89: 482, 499.

[54] Cohen IG (2011) Regulating Reproduction: The Problem with Best Interests. Minnesota Law Review 96: 423, 500.

almost unlimited reproductive liberty.[55] Therefore, while it may not be possible to say that the use of PGT to select for disabling conditions harms *the* future child, the reality is that *a* child is born and that child may encounter the setbacks associated with their disabling condition.

7.4.2 Procreative Beneficence

In the previous chapters, the principle of procreative beneficence was put forward as one of the potential bases for reproductive decision-making in the sphere of embryo testing. Does procreative beneficence support the use of embryo testing for the purposes of selecting for disabling conditions? Savulescu has argued that in the context of deafness, procreative beneficence would not prohibit the selection of an embryo that will become a child who is deaf, if it can be shown that deafness will not adversely impact the welfare of the future child.[56] However, in chap. 4, I argued that disability is essentially negative in nature. Therefore, if a condition will not have a negative impact on the future child's wellbeing, then it will not be considered disabling. On this construction, Savulescu's principle of procreative beneficence may permit a deaf couple, living in the Deaf community, to select an embryo that will become a child who is deaf. However, in the previous chapter, I argued that positive selection should be limited to general-purpose means, characteristics that would be useful to a person, regardless of what kind of life they choose to live, and deafness would not be considered a general-purpose means. In chap. 6 it was noted that limiting positive selection to general purpose means protects the future child's open future. Therefore, if both procreative beneficence and the open future argument are to be foundational principles for reproductive decision-making in the sphere of embryo testing, there needs to be a way to balance them against each other. This will be discussed in the next chapter.

---

[55] The exception is if the future child's life would be so terrible that it would be better for them if they had not been born.

[56] Savulescu J, Kahane G (2009) The Moral Obligation to Create Children with the Best Chance of the Best Life. *Bioethics* 23: 274, 289.

7.5 Non-Medical Sex Selection

This section considers whether prospective parents should have the freedom to select the sex of their future child, using PGT, in the absence of a medical reason.[57] Sex selection for non-medical reasons, sometimes referred to as 'social' sex selection, is currently not permitted in Australia. For example, The NHMRC Guidelines only permit sex for selection for the avoidance of serious medical conditions.[58] Is this prohibition justified, or is it an unacceptable intrusion on the reproductive liberty of prospective parents?

Sex selection for non-medical reasons can take two forms; mere preference and family balancing. Some couples may wish to select the sex of their child due to a preference for one sex over the other. This may be for cultural, religious, or personal reasons. Family balancing, on the other hand, refers to a situation where a couple has a child, or children, of one sex and wish to select a child of the opposite sex to 'balance' their family.[59] Family balancing is based on the idea that bringing up girls (for example) is different to bringing up boys and some parents wish to experience rearing both sexes. It will be argued that non-medical sex selection should not be permitted for either of the purposes outlined above.

7.5.1 *Distortion of the Natural Sex Ratio*

One of the main arguments advanced in opposition to the use of PGT for non-medical sex selection is that it will distort the natural sex ratio. The fear is that this distortion will cause a gender imbalance similar to those seen in China and India. For example, in 2013, China had a birth rate of 117.3 boys for every 100 girls.[60] When compared to a global child sex ratio of 107 boys (aged 0–4 years) per 100 girls,[61] this a cause for concern; particularly in light of China's population of 1.41 billion people at the end of

---

[57] The use of sex selection for the avoidance of a disabling condition is not discussed in this chapter, as the arguments for and against this practice are the same as those pertaining to selection against disabling conditions and are detailed in chap. 5. Some of the arguments opposing PGT for sex selection are arguments against the use of PGT for non-medical reasons. As these arguments were discussed in the previous chapter, they will not be re-considered here.

[58] National Health and Medical Research Council, *Ethical Guidelines on the use of Assisted Reproductive Technology in Clinical Practice and Research 2017*, [8.14].

[59] The term family balancing is curious as it suggests that families with children of only one sex are 'unbalanced'. As Soren Holm notes, 'Who would like to belong to an unbalanced family?' Holm S (2004) Like a Frog in Boiling Water: The Public, the HFEA and Sex Selection. Health Care Analysis 12: 27, 31.

[60] National Bureau of Statistics of China, *Population by Age and Sex (2013)*, China Statistical Yearbook 2014 < http://www.stats.gov.cn/tjsj/ndsj/2014/indexeh.htm > .

[61] United Nations, Department of Economic and Social Affairs, *Sex Ratio by Broad Age Group, Major Area, Region and Country*, World Population Prospects: The 2015 Revision < http://esa.un.org/unpd/wpp/DVD/ > .

2022.[62] It has been suggested that the preference for male offspring is due to China's 'Confucian value system' and former 'one-child policy' that have culminated in a 'one son policy'.[63] Chan and colleagues note that traditionally the majority of Chinese society lived in rural areas and therefore sons capable of physically demanding farm labour were of greater benefit than daughters.[64] However, nowadays less than half the population of China live in rural areas[65] but the 'culture of son preference' continues due to the cultural role played by sons in Chinese society.[66] For example, Daoism and Confucianism, the two major religions in China, require religious family rites to be performed by sons. Further, sons are perceived as lifelong family members, whereas daughters are considered only temporary family members, as once they marry they will join their husband's family.[67] Thus, while sons may be considered an investment due to their obligation to support their parents during old age or ill health, daughters are considered a burden on finite family resources. Even though there may still be a cultural preference for male offspring, recent data from China indicates the child sex ratio is now closer to the global average, with a ratio of 109 boys per 100 girls reported in 2023.[68]

In India, the child sex ratio in 2011 was 108.8 boys per 100 girls. Some Indian states, however, showed a worrying difference in the child sex ratio, with Punjab and Haryana reporting 118.2 and 119.9 boys per 100 girls respectively.[69] In a similar vein to China, the Indian preference for sons stems from the Hindu belief that a son is his parents' pathway to Heaven, as only a son can perform the last rites at his parents' funerals. [70] There is also an economic incentive for India's 'son culture'. Traditionally, the parents of a girl paid a dowry to her husband and his family in exchange for her marriage.[71] Even following the *Dowry Prohibition Act 1961* (India), dowries continue to be paid and dowry payments continue to increase. Therefore, having one or more daughters is a severe financial burden for some Indian parents.

The problem with the natural sex ratio argument is that there is no reason to believe that the distortions prevalent in China and India will also occur in a country

[62] National Bureau of Statistics of China, *Population and Its Composition*, China Statistical Yearbook 2023 < http://www.stats.gov.cn/tjsj/ndsj/2023/indexeh.htm > .

[63] Chan CL, Blyth E, Chan CH (2006) Attitudes to and Practices Regarding Sex Selection in China. Prenatal Diagnosis 26: 610, 610.

[64] Ibid.

[65] The World Bank, *Rural population* https://data.worldbank.org/indicator/SP.RUR.TOTL.ZS.

[66] Chan CL, Blyth E, Chan CH (2006) Attitudes to and Practices Regarding Sex Selection in China. Prenatal Diagnosis 26: 610, 610.

[67] Ibid 618.

[68] National Bureau of Statistics of China, *Population by Age and Gender (2023)*, China Statistical Yearbook 2024 http://www.stats.gov.cn/tjsj/ndsj/2024/indexeh.htm.

[69] Census Organization of India, *Census 2011* http://www.census2011.co.in/sexratio.php.

[70] Dahl E (2003) Procreative Liberty: The Case for Preconception Sex Selection. Reproductive Biomedicine Online 7(4): 380, 383.

[71] Dowries are the transfer of moveable assets from a bride's family to a groom's family. For a detailed discussion of dowries in India see Dalmia S, Lawrence PG (2005) 'The Institution of Dowry in India: Why it Continues to Prevail' The Journal of Developing Areas 38(2): 71.

like Australia. Edgar Dahl argues that to ban sex selection in a western country simply because of problematic practices in other parts of the world is 'wildly inappropriate' and akin to punishing the citizens of one country for the misdeeds of the citizens of another.[72] Further, to cause significant distortions in the sex ratio two conditions need to be satisfied.[73] First, there needs to be a strong preference for offspring of a particular sex. Second, there has to be substantial demand for sex selection using PGT. With regard to the first condition, research conducted by Edith Gray et al. found that the majority of childless couples in Australia do not have a preference for the sex of their first child.[74] Out of those couples who expressed an interest in a child of a particular sex, 35% of male partners and 23% of female partners preferred a boy, compared to just 5.5% of male partners and 16% of female partners stating a preference for their first child to be a girl.[75] This indicates a preference for sons over daughters. However, it has been suggested that in other Western countries there is a preference for the first child to be male,[76] as opposed to an outright preference for sons. Even if there does exist a preference for sons, this falls short of the strong preference required to distort the sex ratio. Further, there would still need to exist a substantial demand for the use of PGT to select for sex.

Rebecca Kippen and her colleagues found that 69% of Australians did not support the use of IVF with PGT for sex selection for non-medical purposes.[77] Moreover, they found that even when couples were asked if pink and blue pills existed that could influence a child's sex, whether these should be legalized, 76% of people did not support legalization of the hypothetical pills.[78] Furthermore, only 7% of the respondents to Kippen's study would have considered taking the pink and blue pills.[79] It is unclear if the number of people interested in taking a hypothetical pill, with unknown side effects, can really tell us anything about the number of people who would undergo a well documented procedure such as IVF with PGT for the purposes of sex selection. Conversely, these results may suggest that if only a very small number of people are willing to take what one may assume would be a relatively cheap and simple pill, the number of people willing to undergo costly and invasive IVF to select the sex of their child would be even smaller. This suggests that demand

---

[72] Dahl E (2003) Procreative Liberty: The Case for Preconception Sex Selection. Reproductive Biomedicine Online 7(4): 380, 383.

[73] Ibid 381.

[74] Gray E, Kippen R, Evans A (2007) A Boy for You and a Girl for Me: Do Men Want Sons and Women Want Daughters? People and Place 15(4): 1, 3.

[75] Ibid.

[76] Dahl E (2007) The 10 Most Common Objections to Sex Selection and Why they are Far from Being Conclusive: A Western Perspective. Reproductive Biomedicine Online 14: 158, 159.

[77] Kippen R, Evans A, Gray E (2011) Australian Attitudes Toward Sex-Selection Technology. Fertility and Sterility 95: 1824, 1825.

[78] For a discussion about the role of public opinion in decisions about ethical issues, see Fovargue S, Bennett R (2016) What Role Should Public Opinion Play in Ethico-legal Decision Making? The Example of Selecting Sex for Non-medical Reasons Using Preimplantation Genetic Diagnosis. Medical Law Review 24(1): 34.

[79] Ibid.

for non-medical sex selection would not be high enough to distort the natural sex ratio in a country like Australia.

Even if there existed conclusive evidence that non-medical sex selection would cause a distortion of the natural sex ratio, it has been suggested that there are other solutions for dealing with this instead of an outright ban. Stephen Wilkinson, for example, suggests a 'pairing system'.[80] Under this system, clinics would be required to pair up couples wanting a boy with couples wanting a girl, so as to ensure no detrimental effect on sex ratios. One would assume that if more couples wanted to select boys than girls, the former would be placed on a waiting list. A pairing system would prima facie appear to be an alternative to an outright ban on non-medical sex selection if the reason for prohibiting this practice was exclusively a concern for the natural sex ratio. However, it will be argued that there are alternative reasons for prohibiting non-medical sex selection, besides concerns relating to sex ratios.

7.5.2 Family Balancing

Some academics have suggested that distortion of the natural sex ratio does not (by itself) provide an adequate justification to stop prospective parents using sex selection for family balancing.[81] However, irrespective of its effect on the natural sex ratio, family balancing may be problematic as it does not benefit the future child,[82] and may actually disadvantage the child by limiting their right to an open future.

Prospective parents claim that they should be permitted to use IVF with PGT for family balancing because they wish to experience rearing both sexes, or because they feel that a child of the opposite sex to the children/child they already have will add an extra dimension to their family. The 'rearing both sexes' argument fails because every child is different and therefore the experience of bringing up each of two children will be different, regardless of their sex, as each child will have their own personality, interests and abilities. This argument and the 'extra dimension' argument, both appear to have gender stereotypes at their core. Therefore, if society permitted non-medical sex selection, it would be sending the message that stereotyping is acceptable and this could result in discriminatory practices. However, as Savulescu notes, it is not acceptable to prohibit non-medical sex selection for reasons of discrimination, because this argument is not accepted in the context of selection against disability.[83]

---

[80] Wilkinson S (2008) Sexism, Sex Selection and "Family Balancing". Medical Law Review 16: 369, 383.

[81] See eg, Dickens BM (2002) Can Sex Selection be Ethically Tolerated? Journal of Medical Ethics 28: 335; Dahl E (2005) Sex Selection: Laissez Faire or Family Balancing? Health Care Analysis 13: 87. In the context of sperm sorting see Robertson JA (2001) Preconception Gender selection. 1(1) The American Journal of Bioethics 1(1): 2.

[82] In *JS and LS v Patient Review Panel* [2011] VCAT 856, the Victorian Civil and Administrative Tribunal refused to grant permission for non-medical sex selection as the procedure was for the benefit of the parents, not the future child.

[83] Savulescu J (1999) Sex Selection: the case for. Medical Journal of Australia 171(7): 373, 375.

Therefore to justify a preclusion on non-medical sex selection, it needs to be shown that there are additional problems with stereotyping in this context that go beyond those identified in the context of selection against disability.

It may be asked, why does family balancing involve stereotyping? Parents who argue that bringing up boys is different to bringing up girls are suggesting that there are fundamental differences between boys and girls that would make the practice of child rearing different, in a way that is somehow different from bringing up two children of the same sex. If sex stereotypes are not at the root of this desire, it must be due to biological differences between the sexes. Perhaps some parents wish to select female children so they can bond over menstrual cycles, pregnancy and eventually menopause.[84] While this cannot be ruled out, it is unlikely that this will be the motivation behind most parents' choice to employ sex selection for family balancing.

A more convincing motivation for wanting a balanced family is that parents want a child of the opposite sex to the ones they already have to satisfy pre-determined gender stereotypes of what bringing up a child of each sex would be like. Berkowitz and Snyder assert that a man's preference for a son with whom he can play 'masculine' sports such as football, stems from a belief that boys are better companions than girls with regard to certain activities.[85] Further, they note that mothers who desire a daughter so they have a shopping companion are basing their decision on the sexist assumption that girls are better at, or perhaps more suited to, shopping than boys.[86] This kind of thinking perpetuates gender stereotypes because unlike, for example, the biological limitations that prevent men from becoming pregnant, no such barriers exist to prevent girls from playing football or boys from shopping.[87] Therefore, Berkowitz and Snyder state that parents wanting to use sex selection 'must have preconceived notions, however vague, about the ramifications of having a certain sexed child: notions which are fundamentally sexist as they are predicated upon anticipated gender based behaviour'.[88] This is where non-medical sex selection becomes particularly problematic.

Imagine a scenario where a family with two sons are permitted to use sex selection to balance their family by implanting an embryo that will become a girl. What if their daughter, like her brothers, enjoys playing contact sports, climbing trees and fishing? Would this be acceptable to her parents? Or would they be disappointed that she does not wear pink, like shopping and play with dolls? Parents who spend large sums of money on IVF with PGT for sex selection may be disappointed if their 'selected'

[84] Wilkinson S (2008) Sexism, Sex Selection and "Family Balancing". Medical Law Review 16: 369, 386.

[85] Berkowitz JM, Snyder JW (1998) Racism and Sexism in Medically Assisted Conception. *Bioethics* 12: 25, 32.

[86] Ibid.

[87] Ibid.

[88] Ibid.

child does not meet the expectations they have for a child of that sex.[89] Further, it is not unthinkable that some parents may push their children, from a young age, into activities stereotypical for their gender. Such behaviour would deprive that child of their right to an open future. Parents may try and limit their child's options to only those they feel are gender appropriate. This may manifest itself in childhood through actions such as forcing a girl to take ballet lessons instead of playing soccer, but it may also have longer lasting ramifications. For example, parents of a daughter may discourage her from pursuing a career in engineering or science as these are traditionally male dominated professions. That girl will therefore have had options closed to her unnecessarily. In a similar vein, a boy who excels at cooking may be steered away from a career in the food industry as his parents may feel that cooking is predominantly women's work. It would be unfortunate for this boy to grow up feeling unable to follow his dream and equally undesirable for him to develop the sexist attitude that certain tasks are only suitable for women. There is therefore a significant risk that the use of non-medical sex selection will deprive future children of their right to an open future.

It is hoped that parents who do not select the sex of their child will be more accepting of the child they get, regardless of whether that child conforms to gender stereotypes. However, it is acknowledged that this will not always be the case. Wilkinson notes that there is a risk that even if the prospective parents are not sexist, the child may find herself growing up in a sexist society and therefore may be more inclined to become a teacher instead of a software developer, because that is what society, or perhaps her peers deem appropriate.[90] If external influences continue to uphold gender stereotypes, what is the harm in permitting parents to select for sex, as they may not be doing any more harm than the societal influences beyond their control? I would argue that just because gender stereotyping exists in society this is not a reason to consider it acceptable, and thus make laws condoning it. If we did, we may be accused of colluding with a sexist society.

7.6 *Human Leukocyte Antigen Typing*

The last part of this chapter considers whether parents should be permitted to use human leukocyte antigen typing (HLA typing) to select a child that has the potential to save the life of an existing sick family member. HLA typing is currently permissible in Australia and in re-considering the regulation of HLA typing, this section also considers whether HLA typing should be permitted in the absence of PGT-M.

[89] Dena Davis refers to this as the 'parental investment factor'. See generally Davis D (2009) The Parental Investment Factor and the Child's Right to an Open Future. Hastings Center Report 39(2): 24–27.

[90] Wilkinson S (2008) Sexism, Sex Selection and "Family Balancing". Medical Law Review 16: 369, 387.

Imagine you are the parent of a seriously ill child. Neither surgery nor medication will cure your child. Your child's only chance of survival is a haematopoietic stem cell transplant. As no suitable donor is available, your only option is to create an embryo through IVF and then subject that embryo to testing to determine if it has the potential to save your existing child. One of these tests is HLA typing and it is undertaken on a pre-implantation embryo to determine whether it is a tissue compatible match with the sick child. This is the situation facing the parents of children suffering from various serious illnesses. If the sick child is suffering from a genetic condition, HLA typing will only be carried out if PGT-M indicates that the embryo does not suffer from the disease affecting the existing child. In the event that the existing child does not suffer from a genetic condition, HLA typing may be used on its own. If the embryo turns out to be a tissue match, it will be implanted into the woman, otherwise it is discarded or donated. Children born following this process are colloquially referred to as 'saviour siblings'. At birth, stem cells are harvested from the saviour sibling's umbilical cord and transplanted into the sick sibling. If the initial transplant is unsuccessful, the younger child may be called upon at a later date to donate bone marrow, or perhaps in rare cases, an organ. It should be noted that HLA typing is only an option for parents whose child can wait over a year for a transplant; as IVF followed by PGT, HLA typing, gestation and birth is a long, uncertain process.[91]

At present, the NHMRC Guidelines advise clinics to seek advice from a clinical ethics committee (or other independent body) before using HLA typing to select a tissue compatible embryo.[92] Further, the guidelines make it clear that HLA typing may only be used to save the life of a sibling, parent, or other relative.[93] This is a change from the previous iteration of the Guidelines which prevented parents from trying to create a tissue compatible embryo for their own benefit or to save the life of a family member who would not be a sibling to the future child.[94] The current guidelines also stipulate that the sick relative must suffer from a 'serious' condition,[95] a change from the life-threatening condition requirement in the previous iteration.[96]

The guidelines suggest that ethics committees should ascertain whether parents considering HLA typing wish to have another child or whether they are simply

[91] Pennings G, Schots R, Liebaers I (2002) Ethical Considerations on Preimplantation Genetic Diagnosis for HLA Typing to Match a Future Child as a Donor of Haematopoietic Stem Cells to a Sibling. *Human Reproduction* 17: 534, 535. Pennings et al. suggest that the entire process could take as long as eighteen months to two years.

[92] National Health and Medical Research Council, *Ethical Guidelines on the use of Assisted Reproductive Technology in Clinical Practice and Research 2017*, [8.17.2].

[93] Ibid [8.17.1].

[94] National Health and Medical Research Council, *Ethical Guidelines on the use of Assisted Reproductive Technology in Clinical Practice and Research 2007*, 55 [12.3].

[95] National Health and Medical Research Council, *Ethical Guidelines on the use of Assisted Reproductive Technology in Clinical Practice and Research 2017*, [8.17.2].

[96] National Health and Medical Research Council, Ethical Guidelines on the use of Assisted Reproductive Technology in Clinical Practice and Research 2007, 55 [12.3.1].

seeking a source of tissue.[97] The commodification of the child as a source of tissue is the most commonly raised objection to the use of HLA typing. Those who raise this objection argue that HLA typing offends the Kantian categorical imperative that a person should not solely be used as a means to another person's ends.[98] A clear contravention of the categorical imperative would arise in the unlikely circumstance where a parent who desperately wants to cure their sick child intends to use the future child's stem cells and then give them up for adoption or simply keep them around as an insurance policy. If, on the other hand, parents wish to have another child and have been delaying doing so due to their existing sick child, HLA typing appears to be a possible solution and would not offend Kant's categorical imperative. As long as the future child is loved as an end in themselves and not just a source of tissue, the categorical imperative does not suggest that HLA typing should be prohibited. Therefore, in a similar vein to the NHMRC Guidelines, a revised framework for embryo testing should contain guidance to clinical ethics committees to carefully consider parents' intentions for having a tissue matched child.

7.6.1 HLA Typing Without PGT

Thus far this discussion has not uncovered anything inherently objectionable about the use of HLA typing. Is the situation different in cases where HLA typing is used in the absence of PGT?[99] If HLA typing without PGT results in a setback to the future child's welfare interests, then there is good reason to prevent parents from using the technology in this way.

As both PGT with HLA typing and HLA typing on its own involve the removal of cells from an early stage embryo, the distinction between them must turn on something other than the testing itself. One possible line of argument is that the difference between the two procedures is that in the first, PGT is initially carried out for the benefit of the future child and therefore tissue typing does not require any further extraction of cells from the embryo. HLA typing on its own, on the other hand, involves the extraction of cells from an embryo solely for the benefit of another person. While HLA typing may not confer a direct benefit on the future child, it may confer on them an indirect benefit. Having a healthy sibling will definitely benefit a child, as will a living, as opposed to a dead sibling. Further, any benefits gained by the parents of a healthy child, such as extra time or resources, will also be passed on to all children in the family.

---

[97] National Health and Medical Research Council, *Ethical Guidelines on the use of Assisted Reproductive Technology in Clinical Practice and Research 2017*, [8.17.2]. It is interesting to note the negative phrasing of the provision, which requires that there be 'no evidence' that the future child will not be welcomed into the family.

[98] Kant I (1898) Critique of Practical Reason and other Works on the Theory of Ethics. Thomas Kingsmill Abbott trans, Longmans, Green and Co., 4th ed, p 52.

[99] HLA typing may be used in the absence of PGT if the condition the sick relative suffers from is not hereditary. One example of such a condition is Diamond-blackfan anaemia.

As the benefit argument does not represent a good reason to draw a distinction between PGT-M with HLA typing and HLA typing on its own, let me turn to wider arguments regarding the welfare of the future child. While concerns about the future child's welfare may give rise to legitimate arguments as to why HLA typing should not be permitted, these arguments tend to apply equally regardless of whether or not PGT-M is used. One such argument centres on the detrimental effect of unsuccessful donation on the psychological wellbeing of the future child. Crystal Liu notes that in such a case, the result will be the same whether or not PGT-M is employed.[100] This is because, whether or not the parents blame the child for the death of the older sibling will more than likely have little to do with the methods employed to secure the child's birth. Liu argues that if psychological harm to the child is a concern, then in order to be ethically consistent a policy should either ban HLA typing in all circumstances, or not draw a distinction between cases where PGT-M and HLA typing are used and cases where HLA typing is used on its own.[101]

The lack of ethical consistency evident in the distinction can also be observed in arguments opposing HLA typing on the grounds that the saviour sibling may grow up in the shadow of their older sibling,[102] or objections centred on human dignity.[103] With regard to the latter, Merle Spriggs argues that it is unclear how a sibling created through the use of HLA typing has had their human dignity compromised, or indeed why the future child's dignity should be considered more important than saving the life of their older sibling.[104] Conversely, Delatycki argues that the human dignity of the saviour sibling needs to be considered, particularly if there is a chance the child will be mistreated if the transplant is unsuccessful and the sick child dies.[105] As stated above, if concerns about the human dignity of the future child are sufficient to prevent parents from accessing HLA typing, this should be so whether or not PGT-M is used. Concerns for human dignity should be addressed by ethics committees considering applications for HLA typing with or without PGT-M. Deciding applications on a case-by-case basis means that clinics can refuse treatment in situations where there is a risk that the future child's human dignity will be compromised.[106]

---

[100] Liu CK (2007) 'Saviour Siblings'? The Distinction between PGD with HLA Tissue Typing and Preimplantation HLA Tissue Typing. Journal of Bioethical Inquiry 4: 65, 69.

[101] Ibid. See also Gavaghan C (2003) Use of Preimplantation Genetic Diagnosis to Produce Tissue Donors: An Irreconcilable Dichotomy? *Human Fertility* 6: 23.

[102] Liu CK (2007) 'Saviour Siblings'? The Distinction between PGD with HLA Tissue Typing and Preimplantation HLA Tissue Typing. Journal of Bioethical Inquiry 4: 65, 69.

[103] Spriggs M (2005) Is Conceiving a Child to Benefit Another Against the Interests of the New Child? Journal of Medical Ethics 31: 341, 341.

[104] Ibid.

[105] Delatycki M (2005) Response to Spriggs: Is Conceiving a Child to Benefit Another Against the Interest of the New Child? Journal of Medical Ethics 31: 343, 343.

[106] Robert Sparrow and David Cram have put forward a controversial alternative solution for treating sick children in need of a stem cell transplant, that does not require any consideration to be given to the welfare of the future child. Sparrow and Cram suggest that where a couple are able to create a disease-free, tissue-matched embryo, but unable to carry a child to term, they should be permitted to create a 'saviour embryo'. They propose extracting and differentiating stem cells from

7.6.2 The Welfare of the Saviour Child

Thus far, I have argued that HLA typing should be permitted when it is carried out with PGT-M as well as in its absence. I now turn to consider the the welfare of the saviour child.

Parents considering HLA typing need to be cognisant of the need to balance the interests of both their existing child and the child they are going to create. It may be argued that Derek Parfit's priority view[107] assists in this balancing act. The priority view would advocate saving the life of the sick child through the use of HLA typing as this child may be deemed worse off than their not yet born sibling. However, this does not mean that the welfare of the saviour sibling is not important. Pennings et al. illustrate the balancing exercise required by the priority view through the analogy of parents who have a child with special educational needs who has to attend a 'special school'.[108] They note that:

> If the parents decide to move in order to be closer to this special school, they harm the other sibling who is losing his friends and familiar environment. This problem cannot be solved in a way that both children benefit. The dilemma for the parents is that they have to balance the interests of both their children.[109]

In a similar vein, HLA typing is permissible because potential harm to the future child can be accepted on the understanding that the treatment will save the life of the existing sick child. However, should the potential harm become real harm, safeguards need to be in place to deal with the psychological trauma or other harm the saviour sibling may experience. If the saviour sibling suffers mental anguish as a consequence of, for example, living in the shadow of their older sibling, their parents and perhaps the state need to work together to alleviate this suffering. Consequently, while HLA typing should be permitted, monitoring and counselling should be provided for the saviour sibling, particularly in the event of an unsuccessful donation. While the NHMRC Guidelines suggest counselling for prospective parents and this has received

---

the embryo to make those cells suitable for transplant: Sparrow R, Cram D (2010) Saviour Embryos? Preimplantation Genetic Diagnosis as a Therapeutic Technology. Reproductive Biomedicine Online 20: 667. This option could also be employed where parents do not want another child, or if there are concerns for the dignity or welfare of the future child. The creation of a saviour embryo would also be an advantage to the sick child, as they would not need to wait over a year for a stem cell transplant.

[107] Parfit D (1997) Equality and Priority. Ratio 10: 202.

[108] Pennings G, Schots R, Liebaers I (2002) Ethical Considerations on Preimplantation Genetic Diagnosis for HLA Typing to Match a Future Child as a Donor of Haematopoietic Stem Cells to a Sibling. *Human Reproduction* 17: 534, 537.

[109] Ibid.

academic support,[110] there appear to be a lack of recommendations for counselling for the saviour sibling, even after academics have noted the potential for harm.[111]

The need for additional safeguards for the welfare of the future child is highlighted by Michelle Taylor-Sands' relational approach to child welfare. Taylor-Sands argues that rather than thinking about the welfare of the child in terms of the best interests of that particular child, a 'more holistic approach' is needed.[112] She contends that as the welfare of the future child is inextricably connected to the welfare of the family into which they will be born, any sound approach to child welfare must recognise 'the importance of connection and interdependence within a family'.[113] Thus, she asserts that a satisfactory account of the welfare of a saviour sibling will be one that includes 'the individual interests of the future child as well as the collective interests the child will share with his/her family'.[114] Unlike children born following PGT for the avoidance of a disabling condition, Taylor-Sands notes that children born as saviour siblings are part of a journey to save the existing child.[115] Therefore, she recommends considering the future child's interests '*in connection with* rather than *in opposition to*, the collective interests the child shares with his/her family'.[116]

Taylor-Sands puts forward a model of regulation that supports prospective parents and emphasises counselling and review by clinical ethics committees. However, as with other regulatory models, the counselling is directed at the parents, pre-conception, with no provision for supporting the parents or the saviour sibling once they are born. With regard to the welfare of the child, she states that 'legislation should specify clear limits on assisted reproductive treatments to protect the welfare of the child to be born'.[117] Further, she supports a 'minimum welfare' approach to treatment, whereby treatment is given 'unless the child born is at risk of exploitation, abuse or neglect'.[118] Consequently, she advocates an approach to welfare that may put the interests of the future child second to those of the family and in particular those of the sick child requiring treatment.[119]

---

[110] See, eg, Taylor-Sands M (2013) Saviour Siblings: A Relational Approach to the Welfare of the Child in Selective Reproduction (Routledge) [6.3.3.2], [6.4].

[111] See, eg, Strong KA et al., 'It's Time to Reframe the Savior Sibling Debate' (2011) 2(3) *AJOB Primary Research* 13, 15; Pennings G, Schots R, Liebaers I (2002) Ethical Considerations on Preimplantation Genetic Diagnosis for HLA Typing to Match a Future Child as a Donor of Haematopoietic Stem Cells to a Sibling. Human Reproduction 17: 534, 537; Liu CK (2007) 'Saviour Siblings'? The Distinction between PGD with HLA Tissue Typing and Preimplantation HLA Tissue Typing. Journal of Bioethical Inquiry 4: 65, 69.

[112] Taylor-Sands M (2013) Saviour Siblings: A Relational Approach to the Welfare of the Child in Selective Reproduction. Routledge, [4.1].

[113] Ibid.

[114] Ibid.

[115] Ibid [4.3.1].

[116] Ibid.

[117] Ibid [6.3.3.3].

[118] Ibid.

[119] Ibid [4.4].

Taylor-Sands' proposal highlights the necessity of support and counselling for the saviour sibling. While many children may find themselves with a sick sibling, the predicament of the saviour sibling, who has been selected specifically for the benefit of their older sibling, is unique. There is thus a clear need for the welfare of saviour siblings to be monitored following their birth, with support and counselling available for them if required. Due to the unique circumstances surrounding their conception, saviour siblings may also benefit from the support of other children and young adults with similar experiences. Taylor-Sands' relational welfare model has been criticised for the view it adopts on familial duty. Colin Gavaghan questions whether this notion justifies the selection of a saviour sibling and as a consequence, if the future child is required to accept these familial obligations purely by virtue of their membership of the family.[120] Gavaghan argues that something more than simply owing a duty to family, that one does not owe to a stranger, is required, however what these additional elements are is not spelt out.[121] Taylor-Sands' response is that additional steps are unnecessary, because familial obligations simply arise out of the relationship itself and 'are grounded in debt arising out of the receipt of benefits and reciprocity created by mutual reliance'.[122]

7.7 Conclusion

This chapter has considered how the concepts of reproductive liberty and the welfare of the future child interact in the context of three uses of embryo testing. I have termed these three uses 'difficult cases' as they raise complex questions about the future child's welfare. The first difficult case was selection for disabling conditions, where a conflict between prospective parents' reproductive liberty and the welfare of the future child is evident. Despite theoretical objections that selection in favour of disability constitutes victimless harm, the conflict should be resolved in favour of the future child. This is because where embryos without a disabling condition exist, these should be implanted so as to minimize harm to the future child, in the sense of not setting back their welfare interests.

The chapter then moved on to sex selection for non-medical reasons and asked whether a prohibition on this practice could be justified by relying on arguments relating to the welfare of the child. It was argued that this use of PGT should not be permitted, as it may disadvantage the future child by limiting their right to an open future and forcing them to conform to gender stereotypes. Lastly, this chapter discussed HLA typing and argued that parents should have the freedom to use HLA

---

[120] Ibid [4.4.1.2]; Gavaghan C (2015) Saviour Siblings: No Avoiding the Hard Questions. Journal of Medical Ethics 41: 931, 931.

[121] Gavaghan C (2015) Saviour Siblings: No Avoiding the Hard Questions. Journal of Medical Ethics 41: 931, 931.

[122] Taylor-Sands M (2015) Saviour Siblings: reply to critics. *Journal of Medical Ethics* 41: 933, 934.

typing either in conjunction with or in the absence of PGT-M. Further, counselling needs to be available to both the saviour child and their parents following the stem cell transplant, including the opportunity for the child to meet others born under similar circumstances. In addition, follow up studies need to be undertaken to examine the effect, if any, that being a saviour child has on that child's wellbeing.

This chapter, together with chaps. 5 and 6, have considered which uses of embryo testing should be permitted in a revised framework for the regulation of PGT. The next chapter looks at the form this regulatory framework should take.

Chapter 8
A Framework for Regulating Embryo Testing in Australia

Abstract This chapter develops a national framework for the regulation of embryo testing in Australia. The proposed framework implements a principles-based model of regulation, while also addressing the need for harmonization and central oversight.

8.1 Introduction

This chapter constructs a national framework for the regulation of embryo testing in Australia. In the previous chapters, I suggested that the principles that should govern embryo testing are firstly, providing the future child with an open future and secondly, permitting parents to act in a procreatively beneficent manner. This chapter adds a third principle to this framework. This third principle would apply only in cases of human leukocyte antigen typing (HLA typing), where a careful balance needs to be struck between the interests of the future child and those of the existing sick relative. This principle therefore requires that in such situations, regard must be had for the welfare of the future child.

The focus of this chapter is on the form that a national framework should take. A key argument in this book is that embryo testing should be regulated at federal rather than state level; this chapter explores why this should be the case. It first summarises the problems with the current state based regulatory scheme, before moving on to explain the advantages of national regulation. The chapter then moves on to consider why principles-based regulation is preferable to rules-based regulation in this sphere. This section also details how a principles-based regulatory framework for embryo testing would operate. The chapter concludes with a proposal for the establishment of a national framework for the regulation of embryo testing, including possible constitutional mechanisms for achieving national regulation, recommendations for a national regulator and enforcement options.

M. de Souza, *The Regulation of Embryo Testing in Australia*,
https://doi.org/10.1007/978-981-96-5902-9_8

8.2 Why Regulate Embryo Testing Nationally?

Three problems with the regulatory framework that currently governs embryo testing will be identified below, and I will argue that these problems can be solved through a system of national regulation. The first problem is that in the states that have enacted legislation, each state has different provisions governing embryo testing. As noted in the preceding chapters, six states have legislation governing assisted reproductive technology (ART) and only three of those have statutory provisions pertaining to embryo testing.[1] These provisions not only differ from each other, but also from the National Health and Medical Research Council's Guidelines (NHMRC Guidelines) which regulate embryo testing in states without legislation. The NHMRC Guidelines provide that clinics may only use preimplantation genetic testing (PGT) to 'select against conditions, diseases or abnormalities that would severely limit the quality of life' of the future child.[2] Clinics are required to comply with the NHMRC Guidelines except where they conflict with a specific legislative provision of the state in which they operate. Having a set of stand-alone guidelines that in some areas duplicate provisions of legislation and in other areas conflict with them seems unnecessary and inefficient. A national framework, depending on its form, would reduce complexity and enable either state or NHMRC resources, perhaps both, to be deployed elsewhere.

In addition to the financial benefits of harmonisation, the need for consistent provisions is important in the event that embryos need to be moved across state borders. Patients need to be confident that embryos created for the purposes of PGT or HLA typing can still have the required tests even if they and thus their embryos need to move to a different state. As PGT becomes more sophisticated and greater options for positive selection arise, having consistent laws across Australia will become of paramount importance.

Harmonisation of the law in this sphere also prevents domestic reproductive tourism, where prospective parents may visit clinics in different states in an attempt to get approval for the test they require. This is particularly undesirable in the context of embryo testing, as time may be critical for prospective parents. For example, imagine a Western Australian couple who wish to use HLA typing to create a 'saviour sibling' for their sick child who has a genetic condition caused by a de novo mutation.[3] The couple (possibly with the ill child) will have to travel interstate to access ART due to restrictions on the use of HLA typing without PGT in Western Australia.[4] This not only delays an already lengthy process, but adds a layer of complexity to an already stressful situation. Therefore, a national framework for the regulation of

---

[1] Victoria, South Australia and Western Australia have legislation governing embryo testing. New South Wales, the Australian Capital Territory and Queensland have ART statutes, but no provisions relating to embryo testing.

[2] National Health and Medical Research Council, Ethical Guidelines on the Use of Assisted Reproductive Technology in Clinical Practice and Research 2017 [8.15.1].

[3] A de novo mutation is a change in the DNA sequence of a gene that has not been inherited.

[4] For details see chap. 2 [2.6.3].

embryo testing would result in better health outcomes for prospective parents and their families.

The second problem with the current regulatory framework is that clinics receive different levels of support depending on the state or territory in which they operate. Western Australia is the only state with a specialist regulator; the Western Australian Reproductive Technology Council (RTC). The RTC takes a hands on approach to the regulation of ART and is involved in the decision-making process for embryo testing. Any conditions for which PGT has not previously been used require the approval of the RTC, which also has a policy on embryo testing that assists clinics with the interpretation of the legislation.[5]

In South Australia, the Ethics and Health Advisory Council (EHAC) was given the task of providing advice on ethical issues arising from ART, following the dissolution of the South Australian Council on Reproductive Technology. However, as the EHAC is by no means a specialist in the field of ART, or even a regulator, its raison d'être was questionable. In 2017, a review of the South Australian regulatory framework by Sonia Allan indicated that clinics would welcome a specialist regulator, or at least an advisory body that could provide specialist advice when required.[6] Further, Allan's report revealed that even though the EHAC was meant to take over some of the functions of the now defunct South Australian Council on Reproductive Technology, this did not in fact happen.[7] At the time of writing EHAC also appears to have ceased operation. These differences between the jurisdictions provide a clear illustration of the different levels of support offered to clinics.

The final problem with the current regulatory framework is the lack of consistency with regard to sanctions for non-compliance. If a clinic located in a state or territory without legislation fails to comply with one of the embryo testing provisions in the NHMRC Guidelines, the clinic may lose their reproductive technology accreditation committee (RTAC) accreditation, however their staff will not be subject to other sanctions.[8] On the other hand, if a clinic in Victoria commits the same breach, for example using PGT for non-medical sex selection, in addition to losing its accreditation, the doctor may also receive a large fine, a term of imprisonment of up to two years, or both.[9] This means that the penalties for the same prohibited activity differ considerably depending on the location of the clinic. A nationwide approach to embryo testing would remove such inconsistencies.

[5] Until recently, Victoria also had a specialist ART regulatory body; the Victorian Assisted Reproductive Treatment Authority (VARTA). On 31 December 2024, VARTA's functions were transferred to the Victorian Department of Health: https://www.varta.org.au. Accessed 22 January 2025.

[6] Allan S (2017) Report on the Review of the Assisted Reproductive Treatment Act 1988 (SA), 27.

[7] Ibid. Allan also noted that in South Australia there was a lack of awareness of the fact that the EHAC even existed.

[8] If the clinic continues to operate after their accreditation has been revoked, the doctor carrying out the procedure may be sentenced to five years imprisonment under the *Research Involving Human Embryos Act 2002* (Cth) s 11.

[9] *Assisted Reproductive Treatment Act 2008* (Vic) s 28(1).

Despite the problems discussed above, some have argued that state based regulation is desirable because it promotes 'innovation and experimentation'[10] and encourages competition between states, with each state competing to make the most innovative laws.[11] This idea of 'interjurisdictional competition'[12] is based on the premise that more effective public policies will attract a greater number of residents (taxpayers) to a state,[13] therefore a state has an incentive to come up with the most innovative laws.[14] However, the interjurisdictional competition argument does not apply in the context of embryo testing, as a couple or individual is unlikely to move to a state on account of their embryo testing policies.[15] Rather, they may engage in reproductive tourism, which as noted above, is not a desirable outcome.[16]

The interjurisdictional competition argument also gives rise to the idea that the states are laboratories in which regulatory experiments are conducted.[17] In the American context, some experiments have been so successful that they have paved the way for national policy making.[18] As Kettl notes, national policy often emerges after federal policy makers observe successful results in state experiments.[19] In the Australian ART context, while some 'laboratories' have been rather busy over the last forty years, others have not. While the reasons for this lack of regulatory activity are unclear, this could be for a number of reasons, including that these states had more pressing issues to deal with or that they felt that inaction was the most appropriate course of action, given the existence of the NHMRC Guidelines.

---

[10] *Gregory v Ashcroft* 501 US 452, 458.

[11] Connell MW (1987) Federalism: Evaluating The Founders' Design. University of Chicago Law Review 54: 1484, 1498. This emphasis on competition, experimentation and creativity has also been noted by Appleby et al.: Appleby G, Aroney N, John T (2012) Australian Federalism – Past, Present and Future Tense. In: Appleby G, Aroney N, John T (eds) The Future of Australian Federalism. Cambridge University Press, p 1–24, p 11.

[12] Kerber W, Eckardt M (2007) Policy Learning in Europe: The Open Method of Co-ordination and Laboratory Federalism. Journal of European Public Policy 14(2): 227, 232.

[13] Connell MW (1987) Federalism: Evaluating The Founders' Design. University of Chicago Law Review 54: 1484, 1498.

[14] This may include not passing any laws. E.g. the decision by the New South Wales government not to legislate on ART until 2007 is one example of inaction being an effective course of action, as clinics in New South Wales were able to offer ART to women who were previously not eligible to access ART in their home state of Victoria. See note 16 below.

[15] The argument is more effective in the context of education policy, where families may move to an area that has a significantly better education system.

[16] This was the situation prior to the decision in *McBain v Victoria* (2000) 99 FCR 116. Single women and lesbians regularly crossed the Victoria–New South Wales border in order to receive treatment. See, eg, G Pennings G (2002) Reproductive Tourism as Moral Pluralism in Motion. Journal of Medical Ethics 28: 337, 337.

[17] This has been termed 'laboratory federalism'. See generally Kerber W, Eckardt M (2007) Policy Learning in Europe: The Open Method of Co-ordination and Laboratory Federalism. Journal of European Public Policy 14(2): 227. In *New State Ice Co v Liebmann* 285 U.S. 262, 52 S. Ct. 371 (1932) Brandeis J referred to states as 'laboratories of democracy'.

[18] Kettl DF (2020) The Divided States of America: Why Federalism Doesn't Work. Princeton University Press, 145.

[19] Ibid 147.

8.2.1 Advantages of National Regulation

National regulation brings with it the possibility of a national regulator, like the Human Fertilisation and Embryology Authority (HFEA) in the UK (discussed in detail in chap. 3). This would provide all clinics, regardless of where they operate, with the same level of support with regard to legal and ethical issues. A national regulator may also be more cost effective than separate state regulators.

Two reviews undertaken by the UK government illustrate the importance of a national regulator for ART. These reviews formed part of the UK government's 2010 decision to reduce the number of arm's-length bodies[20] that existed.[21] In June 2012, a government consultation was launched seeking public opinion on the desirability and indeed feasibility of transferring the HFEA's functions to the Care Quality Commission and Health Research Authority.[22] Seventy five per cent of respondents opposed this transfer.[23] One of the most common reasons given was that the HFEA had 'developed considerable expertise in highly specialised fields'.[24] As a result, they were considered 'trusted and respected by the regulated sectors'.[25] As one respondent remarked, '[r]egulation in this controversial sector would not be adequate in the absence of a specialist regulatory body'.[26] Respondents were also asked whether the HFEA's functions should be transferred to bodies other than the Care Quality Commission. While some respondents suggested transferring policy making functions to the UK's Department of Health, others noted the importance of having policy functions separate from government and thus 'free from political interference'.[27]

The HFEA's knowledge and expertise in a constantly evolving area, and the fear that this could be lost should functions be transferred elsewhere played a significant role in the reasoning of those who supported keeping the HFEA.[28] Further, having a single regulator that specialises in ART was seen as preferable to having the regulation

[20] The UK's National Audit Office states that the term arm's-length body is an umbrella term that covers a number of different types of bodies including non-ministerial departments and non-departmental public bodies: National Audit Office, *Departments' Oversight of Arm's-length Bodies: A Comparative Study*, House of Commons Paper No 507, Session 2015–2016 (2016) 5 [1].

[21] *Public Bodies Act 2011* (UK). Under this legislation, various bodies were targeted for abolition, merger or to have their functions modified or transferred to a different body. In particular, the number of health-related arm's-length bodies was seen as excessive. Among the bodies that were considered for modification or to have their functions transferred elsewhere was the HFEA.

[22] Department of Health (UK), *Consultation on Proposals to Transfer Functions from the Human Fertilisation and Embryology Authority and the Human Tissue Authority* (2012).

[23] Department of Health (UK), *Government Response to the Consultation on Proposals to Transfer Functions from the Human Fertilisation and Embryology Authority and the Human Tissue Authority* (2013) 7 [9].

[24] Ibid [10].

[25] Ibid.

[26] Ibid 13 [37].

[27] Ibid 29 [100].

[28] Ibid 31–32.

of ART spread across multiple organisations.[29] Another reason cited by respondents for retaining the HFEA was that it had built a brand and the public had confidence in that brand. As one stakeholder noted, the 'HFEA provides a sound regulatory framework that is recognised internationally. The organisation has public confidence in a sensitive area of practice'.[30]

Similar themes arose when the UK government considered merging the HFEA with the Human Tissue Authority.[31] The importance of an independent regulator with a focus on and specific expertise in ART was seen as crucial for the achievement of effective regulation.[32] This can be contrasted with South Australia's unsuccessful 'experiment' with a generalist healthcare ethics body (discussed earlier in this chapter). The South Australian experience together with the laissez faire approach of a number of states, serves to reinforce the argument that Australia needs a national regulator that specialises in ART.[33]

The UK government's reviews of the HFEA can teach Australia a number of lessons. Most importantly, they highlight the value of a national regulator. The fact that three quarters of the respondents to the UK government's consultation supported the HFEA's existence, indicates the value of a dedicated regulatory body in this sector. Ideally, the regulator needs to be independent from the government and should specialise in ART, rather than having regulatory oversight of multiple areas. The independence of the proposed regulator is important because such a regulator will need to deal with sensitive issues and should be free to do so without government interference.[34] Further, the new regulator should be tasked with 'building a brand' around its activities in order to gain public confidence and the respect of key stakeholders.

The desirability of national regulation is not unique to embryo testing or even ART generally. Calls for a national regulatory framework can also be seen in other

---

[29] Ibid 32 [109].

[30] Ibid [110].

[31] McCracken J (2013) *Review of the Human Fertilisation & Embryology Authority and the Human Tissue Authority*. Available via Department of Health and Social Care (UK). https://www.gov.uk/government/publications/review-of-human-fertilisation-embryology-authority-and-human-tissue-authority. Accessed 23 January 2025.

[32] Department of Health (UK), *Response to the Review of the HFEA and HTA: Government Response to the Independent Review of the Human Fertilisation and Embryology Authority and the Human Tissue Authority by Justin McCracken* (2013) 9.

[33] This argument carries even more weight if one looks outside the sphere of embryo testing to the broader ART landscape. In 2024, the ART industry in Queensland generated headlines across Australia due to inadequate regulation. See, eg Meacham S (2024) Incest Fears After Sperm Donor's 200-plus Visits. The Canberra Times. https://www.canberratimes.com.au/story/8684244/incest-fears-after-sperm-donors-200-plus-visits/.

[34] See Department of Health (UK) (2010) Liberating the NHS: Report of the Arm's-length Bodies Review [3.29].

areas such as embryo research,[35] privacy[36] and even tort law.[37] For example, in its report, *For Your Information: Australian Privacy Law and Practice*, the Australian Law Reform Commission found that there was strong support for nationally consistent privacy laws. Among the problems identified by the Australian Law Reform Commission were the inconsistency of the legislative provisions that applied in each state or territory,[38] and related to this, the difficulties experienced when trying to share information across state borders[39] and the confusion encountered by people wishing to make a privacy complaint.[40] They concluded that a 'nationally consistent privacy regime would lessen unjustified compliance burdens and cost'.[41]

The issue of decreasing the compliance burden and associated costs noted by the Australian Law Reform Commission can also be applied to embryo testing.[42] Instead of requiring clinics to comply with their obligations under the relevant statute, the NHMRC Guidelines and the RTAC Code of Practice, the compliance burden on clinics would be drastically reduced through the establishment of one national regulatory body, with its own code of practice and corresponding legislation. Similar benefits could be gained in terms of costs, as state based regulators, such as Western Australia's Reproductive Technology Council could be dissolved and expertise transferred to the newly formed national regulator.

The benefits of national regulation extend beyond the establishment of a national regulator. Consistent laws across Australia should make it easier for both clinics and patients to identify their obligations and understand their rights and options. National regulation will also result in the same penalties for non-compliance across Australia. These penalties may be set out in legislation and if required can be applied by the regulator.

As noted in chapter 1, one of the advantages of regulating embryo testing is that it allows for the collection and dissemination of data and other information.[43] National regulation would bring with it the opportunity to make publicly available

[35] *Research Involving Human Embryos Act 2002* (Cth). See also Legislation Review Committee (2005) Parliament of Australia, *Prohibition of Human Cloning Act 2002 and the Research Involving Human Embryos Act 2002* 162 [17.2].

[36] Australian Law Reform Commission (2008) For Your Information: Australian Privacy Law and Practice, Report No 108 vol 1, 192 [3.9].

[37] Panel of Eminent Persons (2002) Review of the Law of Negligence: Final Report, 26 [1.9].

[38] Australian Law Reform Commission (2008) For Your Information: Australian Privacy Law and Practice, Report No 108 vol 1, 192 [3.9].

[39] Ibid.

[40] Ibid 193 [3.13].

[41] Ibid 192 [3.9].

[42] The Fertility Society of Australia and New Zealand has suggested that decreasing the compliance burden on clinics may lead to a reduction in the cost of ART in general. See Fertility Society of Australia and New Zealand (2024) Draft Findings, Recommendations and Framework for FSANZ 10 Year National Fertility Roadmap, 7. Accessible via https://www.fertilitysociety.com.au/about/media-news/ Accessed 30 January 2025.

[43] Information regarding the tests carried out on embryos and the success rates of embryo testing and other ART was previously available through the VARTA website, but this data only related to clinics in Victoria.

information about the tests carried out on embryos and the success rates of PGT in respect of clinics throughout Australia. This kind of transparency is desirable because it allows prospective parents to make an informed choice. In 2021, the Australian Government launched Your IVF Success, a website aimed at increasing transparency around the success rates of ART clinics.[44] While the website is a step in the right direction, it does not provide any information about PGT. Consequently, the websites of individual clinics are the main source of information about embryo testing. This information should be made available through an independent regulator who does not stand to make financial gains from such testing, as was the case with the VARTA.[45]

The foregoing discussion highlights the importance of having a national regulator for ART. While the focus of this book is embryo testing, it is hoped that the following discussion will serve as a starting point from which to build a national framework not just for embryo testing, but also for ART generally.

8.3 Principles-Based Regulation

Having made the case for a national regulatory framework for embryo testing, this chapter now considers the substantive content of the proposed system of regulation. Building on the principles that were put forward in the preceding chapters, this section will focus on the concept of principles-based regulation. Principles-based regulation refers to a type of regulation that principally relies on broad principles, as opposed to detailed rules.[46] This section will firstly look at what is meant by the term 'principles' and how principles differ from rules. Consideration will then be given to the question of what principles-based regulation is and how it works. Lastly, two questions will be asked. First, how would principles-based regulation work in the context of embryo testing? Second, what are the disadvantages of using a principles-based framework to regulate embryo testing?

The first question that needs to be asked is, what is the difference between a principle and a rule? This distinction can be illustrated by using a simple example. A rule may state that 'drivers must not exceed the fifty kilometres per hour speed limit'. A principle, on the other hand, may provide that 'drivers of motor vehicles must drive at a reasonable speed at all times, in order to avoid collisions with other road users'. As Black notes, one of the features of a principle is the lack of quantitative terms and the inclusion of an evaluative aspect.[47] Unlike the rule in the above example that contains a quantitative limit, the principle requires drivers to make a determination

[44] Your IVF Success https://yourivfsuccess.com.au Accessed 4 February 2025.

[45] See Victorian Assisted Reproductive Treatment Authority, *Pre-implantation Genetic Testing* https://www.varta.org.au/resources/information-sheets/pre-implantation-genetic-testing-pgt Accessed 24 January 2025.

[46] Black J, Hopper M, Band C (2007) Making a Success of Principles-based Regulation. Law and Financial Markets Review 1: 191, 191.

[47] Ibid 192.

with regard to what they perceive to be a reasonable speed. The above example also illustrates a further distinction between rules and principles; that rules apply in an 'all-or-nothing fashion'.[48] In the above example, a person is either in excess of the prescribed speed limit, or they are not. Therefore, speed limits can easily be classified as rules. Another difference between principles and rules that is evident from this example is that principles are written so as to highlight the reasoning behind a provision.[49] In this example, the principle explains why drivers must drive safely, the rule, on the other hand, contains no such explanation. Principles-based regulation is not appropriate in all circumstances, as the above example illustrates. In some contexts, such as road safety, rules are a more appropriate method of regulation.

Principles may be described as 'broad-based' standards[50] that express fundamental objectives. However, this description is not accepted by everyone. Braithwaite suggests that principles are 'unspecific or vague prescriptions'[51] and similarly Raz notes that principles 'prescribe highly unspecific actions'.[52] Even though both Braithwaite and Raz describe principles as being prescriptive in nature, Burgemeestre et al. note that 'principles do not prescribe, but they point in a certain direction'.[53] As a result, it is possible for two principles to point in different directions and in any regulatory framework based on principles it must be possible to ascertain which principle carries the most weight.[54]

Principles 'are drafted at a high level of generality', thus making them flexible enough to keep pace with rapidly changing industries or technologies.[55] They are also broadly applicable to a wide range of scenarios.[56] The successful application of principles also requires more knowledge of the subject matter and context than is required in the application of rules.[57] A further characteristic of principles is that they specify outcomes, but leave the finer details of the decision-making process, or method of achieving the outcome, to the relevant party.[58] For this reason, principles

---

48 Dworkin RM (1967) The Model of Rules. The University of Chicago Law Review 35(1): 14–46, 25.

49 Black J, Hopper M, Band C (2007) Making a Success of Principles-based Regulation. Law and Financial Markets Review 1: 191, 192.

50 Ibid.

51 Braithwaite J (2002) Rules and Principles: A Theory of Legal Certainty. Australian Journal of Legal Philosophy 27: 47, 47.

52 Raz J (1971) Legal Principles and the Limits of Law. Yale Law Journal 81: 823, 838.

53 Burgemeestre B, Hulstijn J, Tan Y-H (2009) Rule-based versus Principle-based Regulatory Compliance. Frontiers in Artificial Intelligence Applications 205: 37, 39.

54 Ibid.

55 Black J, Hopper M, Band C (2007) Making a Success of Principles-based Regulation. Law and Financial Markets Review 1: 191, 192.

56 Ibid.

57 Burgemeestre B, Hulstijn J, Tan Y-H (2009) Rule-based versus Principle-based Regulatory Compliance. Frontiers in Artificial Intelligence Applications 205: 37, 39.

58 Gunningham N (1996) From Compliance to Best Practice in OHS: The roles of specification, performance and systems-based standards. Australian Journal of Labour Law 9(3): 221, 222.

Table 8.1 Summary of the differences between principles and rules

Principles	Rules
Focus on outcome	Focus on outcome and method
Apply broadly	Apply narrowly
Require interpretation	Minimal/no interpretation required
Knowledge of subject and context required for successful application	Minimal knowledge of subject and context required
Flexible—can keep pace with technological and industrial changes	Rigid—require re-drafting in order to stay valid
Highlight the reasoning behind the provision	Do not necessarily explain why a specific action is required
Compliance may occur to a certain extent	Apply in an all-or-nothing fashion
Unlikely to contain quantitative terms	May contain quantitative terms
Contain an evaluative aspect	Usually do not contain evaluative terms

are sometimes referred to as performance standards and thus principles-based regulation as performance-based regulation.[59] A rule, on the other hand, instructs a party to take particular measures in order to achieve a specific outcome.[60] As a consequence, it requires little interpretation from the party at whom it is directed.[61] As minimal interpretation is required, the level of knowledge required of the person applying the rule is significantly less than that required for the application of principles. Table 8.1.

8.3.1 Principles-Based Regulation in Practice

Having identified the differences between principles and rules, this section considers how principles-based regulation works in practice, using the UK financial services model as an example. The key feature of principles-based regulation is the reliance on principles, as opposed to detailed rules, to guide decision-making. Rules may play a part in principles-based frameworks, but their role will be secondary to the role of principles.[62]

In the UK, the Financial Conduct Authority (FCA) operates within the principles-based regulatory framework originally developed by its predecessor, the Financial Services Authority.[63] This model of principles-based regulation is worthy of consideration as the Financial Services Authority's original model has been widely

[59] Ibid.

[60] Ibid.

[61] Ibid.

[62] Black J (2010) The Rise, Fall and Fate of Principles-based Regulation. In: Alexander K, Moloney N (eds), Law Reform and Financial Markets. Edward Elgar, 3, 7.

[63] Financial Conduct Authority (2013) FCA Handbook. https://www.handbook.fca.org.uk Accessed 24 January 2025.

discussed in academic literature[64] and prior to its dissolution, the Financial Services Authority had been described as a 'thoughtful leader on principles-based financial services regulation'.[65] The FCA sets out eight principles which it refers to as the 'principles of good regulation',[66] in addition to the twelve principles which businesses must follow. The FCA's 'principles of good regulation', lay out the overarching themes inherent in the regulatory framework. These include the principles of 'efficiency and economy',[67] in which the FCA commits to using its resources in an efficient and economical manner; 'proportionality',[68] whereby the FCA undertakes only to impose restrictions on businesses that are proportionate to the expected benefits to be gained from those restrictions, and 'sustainable growth',[69] highlighting the FCA's commitment to contribute to the medium or long-term growth of the UK economy.[70]

In addition to the above eight principles, the FCA describes its principles for businesses as 'a general statement of the fundamental obligations of firms under the regulatory system'.[71] They have also been referred to as 'a high-level code of ethics for financial services'.[72] As mentioned above, there are twelve principles that businesses regulated by the FCA must adhere to; these include principles related to integrity, 'skill, care and diligence', communication with clients and conflicts of interest.[73]

Failure to comply with any of the twelve principles may result in disciplinary action against the offending business.[74] The burden of proof is on the FCA to show

---

[64] See eg, Black J, Hopper M, Band C (2007) Making a Success of Principles-based Regulation. Law and Financial Markets Review 1: 191; Black J (2010) The Rise, Fall and Fate of Principles-based Regulation. In: Alexander K, Moloney N (eds), Law Reform and Financial Markets. Edward Elgar, 3; Black J (2008) Forms and Paradoxes of Principles-based Regulation. Capital Markets Law Journal 3: 425; Schwarcz SL (2009) The 'Principles' Paradox. European Business Organization Law Review 10(2): 175; Awrey D (2010) Regulating Financial Innovation: A More Principles-based Proposal. Brooklyn Journal of Corporate Financial & Commercial Law 5: 273.

[65] Ford CL (2008) New Governance, Compliance, and Principles-Based Securities Regulation. American Business Law Journal 45(1): 1, 14.

[66] Financial Conduct Authority, *Principles of Good Regulation* (7 July 2016) < https://www.fca.org.uk/about/principles-good-regulation > .

[67] Ibid principle 1.

[68] Ibid principle 2.

[69] Ibid principle 3.

[70] The other principles are 'consumer responsibility', 'senior management responsibility', 'recognising the differences between the business carried on by different regulated persons', 'openness and disclosure' and lastly, 'transparency'.

[71] Financial Conduct Authority (2013) FCA Handbook, PRIN 1.1.2. https://www.handbook.fca.org.uk Accessed 24 January 2025.

[72] MacNeil I (2015) Rethinking Conduct Regulation. Butterworths Journal of International Banking and Financial Law 30: 413, 417.

[73] Financial Conduct Authority (2025) FCA Handbook, PRIN 2.1.1. https://www.handbook.fca.org.uk Accessed 11 August 2025.

[74] Ibid PRIN 1.1.7.

that a firm has breached one or more of the principles.[75] The FCA has noted that businesses are expected to comply with the 'spirit' of the provisions and take into account the intention behind the provisions.[76] Further, a business can breach the principles even when they have not broken the law. For example, in 2013, the FCA fined two banks for breaching the third principle, that of management and control. The FCA imposed the fines on the basis that neither bank took reasonable care in the implementation of their anti-money laundering policies.[77] In March 2018, Vanquis Bank were fined for breaching both the sixth and seventh principles.[78] The FCA found that Vanquis did not disclose all the relevant information to customers about a product they were selling.[79] The FCA concluded that Vanquis had not treated their customers fairly, thus breaching principle six and further, by not communicating the relevant information to them, had also breached principle seven.

The problem with the FCA's principles-based regulatory framework is that the principles for businesses act as an extra layer of regulation in addition to the plethora of statutes and regulations that businesses must follow. It has been suggested that while the twelve principles may provide 'some high-level guidance as to the ethical objectives of conduct regulation, there remains a disconnection with the underlying rules'.[80] Thus in the development of a principles-based framework, one lesson that can be learnt from the FCA is to ensure that the principles are developed in connection with any other rules that exist in the field.

8.3.2 Principles-Based Regulation of Embryo Testing

Having discussed principles-based regulation in a general sense, this section considers how principles-based regulation applies to the field of embryo testing.

---

[75] Ibid.

[76] Financial Conduct Authority (2014) Detailed Rules for the FCA Regime for Consumer Credit, 83 [11.6].

[77] Financial Conduct Authority, *FCA Fines EFG Private Bank £4.2 m for Failures in its Anti-money Laundering Controls* (13 May 2013) < https://www.fca.org.uk/news/press-releases/fca-fines-efg-private-bank-£42m-failures-its-anti-money-laundering-controls > ; Financial Conduct Authority, *FCA fines Guaranty Trust Bank (UK) Ltd £525,000 for Failures in its Anti-money Laundering Controls* (17 September 2013) < https://www.fca.org.uk/news/press-releases/fca-fines-guaranty-trust-bank-uk-ltd-£525000-failures-its-anti-money-laundering > . In June 2018 principle three was breached by Canara Bank for similar reasons. Financial Conduct Authority, *FCA fines and imposes a restriction on Canara Bank for anti-money laundering systems failings* (6 June 2018) < https://www.fca.org.uk/news/press-releases/fca-fines-and-imposes-restriction-canara-bank-anti-money-laundering-systems-failings > .

[78] Financial Conduct Authority, *The FCA Fines Vanquis £1,976,000 and Orders Vanquis to Pay Compensation to Customers* (27 February 2018) < https://www.fca.org.uk/news/press-releases/fca-fines-vanquis-1976000-and-orders-vanquis-pay-compensation-customers > .

[79] Ibid.

[80] MacNeil I (2015) Rethinking Conduct Regulation. Butterworths Journal of International Banking and Financial Law 30: 413, 417.

Principles-based regulation is an appropriate method for the regulation of embryo testing for two reasons. First, principles-based regulation is inherently more capable of keeping pace with innovation than rules-based regulation.[81] As the principles at the heart of a principles-based framework are broad, they can easily adapt as new developments in embryo testing are discovered. Therefore, principles do not become as easily out dated as can be the case with rules, and thus do not need constant amendment. This is not only financially desirable, as it saves the considerable resources required to update rules, but also means that new uses of embryo testing do not get held up due to the bureaucracy involved in amending legislation.

The second reason principles-based regulation would work well in the sphere of embryo testing is because principles are more flexible than rules. This flexibility not only allows principles to keep pace with innovation, as noted above, but also gives the decision-making process an element of flexibility. Thus the principles should be broad enough to support and promote a consultative joint decision-making process between clinicians and their patients, with guidance from the regulator if required. The flexibility provided by principles-based regulation also avoids the pitfalls of interpreting prescriptive rules, for example, having to decide whether a genetic condition meets the threshold of seriousness as required by the South Australian[82] and Western Australian Acts.[83] Clinics will instead have the discretion to conduct any tests they feel are appropriate, so long as they comply with the overarching objective of the principles. A principles-based regulatory framework would also remove prescriptive requirements from the NHMRC Guidelines, for example, the requirement that testing may only be undertaken for conditions that would 'severely limit the quality of life' of the future child.[84] Principles are also desirable because they explain the rationale behind any restrictions on testing that are imposed.

This book argues that there are three principles on which the regulation of embryo testing should be based. The first principle is that embryo testing and the subsequent selection of embryos should preserve the future child's right to an open future. This may be referred to as the 'protective principle', because it protects the future child from decisions that may be detrimental to their right to an open future. As discussed in chaps. 5 to 7, a child's right to an open future is a right to self-determination and parents should not make decisions that would give a future child fewer options than they could otherwise have had. The second principle is that parents should be permitted to act in a procreatively beneficent manner (the beneficence principle). As discussed in the earlier chapters, procreative beneficence is the idea that prospective parents have good reasons to choose the embryo that is expected to have the best life. This principle gives prospective parents the autonomy to make decisions about what they feel would be the best life for their future child. Since principles may

---

[81] Devaney S (2011) Regulate to Innovate: Principles-based Regulation of Stem Cell Research. Medical Law International 11: 53, 57.

[82] Assisted Reproductive Treatment Act 1988 (SA) s 9(c)(iii).

[83] *Human Reproductive Technology Act 1991* (WA) s 14(2b)(a)(ii).

[84] National Health and Medical Research Council, *Ethical Guidelines on the Use of Assisted Reproductive Technology in Clinical Practice and Research 2017* [8.16].

point in different directions, there must be a way to reconcile a conflict between the two principles. In the proposed framework, if the principles conflict, principle one overrides principle two. This means that a child's right to an open future is given greater weight than the freedom of prospective parents to act in a procreatively beneficent manner. The third principle is specific to HLA typing and takes into account the reality that the interests of the future child need to be weighed against the needs of the existing sick relative (the balance principle).

These principles are broad enough to give parents the freedom to request tests beyond those for serious genetic conditions. In chap. 5, I argued that there was no sound reason to limit embryo testing to serious genetic conditions and therefore prospective parents should have the freedom to test for and select against any condition that could be described as disabling. Under the proposed framework, this kind of testing could be authorised under either of the principles. Selecting against a disabling condition protects the future child's right to an open future and is also one way in which parents can act in a procreatively beneficent manner.

In chap. 6, I argued that PGT for polygenic conditions should be permitted once more research has been conducted and regulatory processes put into place. Should the technology become commercially available, prospective parents should also have the freedom to use PGT for the selection of polygenic traits, provided that the characteristic or trait under consideration is also a general-purpose means. This is one area in which principles-based regulation will help the law keep up with scientific innovation. Instead of updating the law each time a new kind of test becomes available, a principles-based regulatory framework would allow clinics to assess requests for PGT based on whether they fall within the principles.

While in some cases concerning the selection of polygenic traits application of either principle will lead to the same result, in other cases, applying each principle separately may result in conflicting conclusions. Consider, for example, prospective parents who would like to test their embryos for educational achievement, as they believe that a child with the genes that code for high educational attainment[85] will have a better life than a child without those genes. However, once the embryos are tested, the one that scores highest for educational attainment also has a high polygenic score for bipolar disorder.[86] The prospective parents nevertheless wish to select this embryo. While these parents may believe that they are acting in a procreatively beneficent manner, this choice may run contrary to the first principle. Therefore, even though the attributes that correspond to high educational attainment may be general-purpose means, the situation is more complex due to the increased risk of bipolar disorder. Consequently, prospective parents will need to discuss such a case with their reproductive healthcare team to come to a decision about whether bipolar

[85] Lee et al. note that education attainment is 'moderately heritable'. See Lee JJ, Wedow R, Okbay A et al. (2018) Gene Discovery and Polygenic Prediction from a Genome-wide Association Study of Educational Attainment in 1.1 million Individuals. Nat Genet 50: 1112–21.

[86] Turley et al. note that embryos with high polygenic scores for educational attainment, also showed an increased risk of bipolar disorder. Turley P, Meyer M, Wang N et al. (2021) Problems with Using Polygenic Scores to Select Embryos. N Engl J Med 385(1): 78–86, 81.

disorder would setback a future child's interests, or whether it is compatible with them having an open future.

In chap. 7, I argued that the prohibition on selection for disabling conditions should continue. This ban could continue under the proposed framework through the application of the principles. To use an oft cited example, a deaf couple may want to test their embryos to find out if any of them have inherited the mutation that would result in the birth of a deaf child. They may argue that the principle of procreative beneficence gives them the freedom to test for deafness, as a deaf child will more easily fit into the Deaf culture in which they live than a hearing child. On this basis, they may reason that a deaf child will have a better life than a hearing child. However, as principle one overrides principle two, consideration must also be given to the future child's open future. As noted in chap. 7, while being unable to hear may not be a disadvantage in the Deaf community, it becomes a setback if a person chooses to leave the Deaf community. This is supported by the interactional model of disability, which was the model selected in chap. 4 as the most appropriate for reproductive decision-making. Thus all things being equal, having opportunities beyond those that may be available in the Deaf community is preferable to having limitations placed on one's opportunities prior to being born. Consequently, foreclosing the opportunities offered by being able to hear before the child is even born, may offend the protective principle.

The second type of embryo testing considered in chap. 7 was non-medical sex selection. The current prohibition on this kind of testing would continue under the proposed framework, as the future child's right to an open future under the protective principle will override parental desires for a child of a specific sex. As non-medical sex selection may diminish a child's right to an open future by forcing them to embrace gender stereotypes from an early age, this kind of selection would not comply with the protective principle.

The final kind of embryo testing considered in this book was HLA typing. HLA typing may be carried out with or without PGT, depending on the condition for which a tissue match is required. It was argued in chap. 6 that HLA typing should potentially be permitted in either circumstance. This kind of testing presents an anomaly, as it is difficult to argue that being a 'saviour sibling' or 'saviour relative' advances a child's right to an open future. However, provided that the prospective parents genuinely want a second child and not merely a source of tissue, it is possible to argue that in the absence of a negative impact on the future child's right to an open future, HLA typing may be justified on the basis of the significant benefit gained by the older sibling or other relative. This begs the question, how should HLA typing be regulated under the proposed framework? Three possible options are available; first, are there any benefits to the future child from HLA typing that could be said to advance their right to an open future? In the alternative, should HLA typing be a written exception to the application of the principles? Last, should an additional principle be developed to address HLA typing?

Considering the above three options in turn, it may be argued that HLA typing promotes the future child's right to an open future by giving them the opportunity to grow up in a healthy family. While this may be a tenuous link, it is likely that the additional time and money the parents will have as a result of no longer caring for a sick child will benefit all the children of the family and it is in this way that the future child's interests are advanced. Similar reasoning could be used to justify HLA typing as a procreatively beneficent choice. Prospective parents may argue that a child growing up with a healthy sibling will have a better life than a non-HLA matched child born into the same family, as the latter may have to endure the death of a sibling at an early age. The problem with these arguments is that an HLA matched sibling will not always be able to cure the sick older child and therefore if this argument is to be accepted the converse proposition should also be worthy of consideration. If a stem cell transplant is not successful, the parents' priority will be finding a cure for the sick child and the saviour sibling may become neglected and thus deprived of their right to an open future. Therefore, rather than trying to authorise HLA typing by arguing that it complies with the protective principle, the other two regulatory options should be considered.

The second regulatory option is that a written exception to the protective principle is introduced into the framework, in order for HLA typing to be undertaken. A limited exception, perhaps included in explanatory notes, may read as follows: 'In circumstances where embryo testing is required for the purposes of human leukocyte antigen typing, in order to save the life of an individual who will be a family member of the future child, compliance with the principles is not required.' The exception could include further details about the alternative considerations that should be taken into account, such as a requirement to consider the welfare of the future child. The proposed framework would also include the recommendations outlined in chap. 7 for ongoing monitoring and support of the future child. The benefit of a limited exception is that it avoids the need to over-stretch concepts such as the open future argument and procreative beneficence. A possible concern with this approach is that it may lead to a slippery slope, where other kinds of tests that do not comply with the principles are also permitted.

The third possible regulatory option for HLA typing is to create an additional principle that specifically deals with this kind of testing. This principle may provide that: 'in cases where embryo testing is utilised for the purposes of creating a tissue matched child, in order to save the life of a person who will be a family member of the future child, regard must be had for the welfare of the future child.' Guidance accompanying the principles could clarify that compliance with principle one is not mandatory in cases of HLA typing, however, alternative welfare considerations must be assessed, including those pertaining to the ongoing welfare of the future child. The inclusion of an additional principle has the benefit of simplicity, in that it does not require the use of convoluted reasoning or exceptions to create the desired result. While a principle that governs a specific kind of testing is considerably narrower than the other principles, the proposed principle still provides flexibility for decision-makers.

Having considered three possible regulatory options for the inclusion of HLA typing in the proposed framework, I suggest that the most suitable method is the introduction of an additional principle – the balance principle.

8.3.3 Disadvantages of Principles-Based Regulation

Before moving on to consider the form which a national principles-based framework may take and the mechanism through which this may be achieved, it is worth considering the disadvantages of adopting a principles-based regulatory framework. The main challenge of principles-based regulation has been referred to as the 'principles paradox'.[87] Julia Black, for example, criticises principles-based regulation for failing to provide 'certainty and predictability'.[88] Black notes that certainty and predictability are distinct concepts.[89] She asserts that certainty refers to a common understanding about the meaning of a principle and how it applies. Predictability, on the other hand, refers to the foreseeability of a regulator's actions.[90] This is important as regulated entities need to be confident about the response they will receive from the regulator. As noted above, the aim of the principles in principles-based regulation is to provide flexibility to decision-makers and also ensure that the law keeps pace with innovation. Unfortunately, due to the principles paradox, the converse result sometimes transpires.[91] The interpretation of principles may become overly conservative, more so than they would have been under detailed rules. This is often due to regulated entities becoming concerned about breaching the principles and therefore acting in an overly cautious manner.

One way to overcome this paradox is through the existence of a regulator that provides guidance, if required, during the decision-making process. Schwarcz suggests that 'interpretive guidelines' may also assist in resolving the paradox.[92] In constructing a framework for embryo testing, it is important to ensure the principles paradox is avoided. As the purpose of adopting a principles-based framework is to give prospective parents the freedom to utilise new applications of PGT, the principles paradox is the very antithesis of what the proposed regulatory framework aims to achieve. The principles paradox may be overcome through the establishment of a national regulator for embryo testing that has an advisory function. As part of the regulator's advisory function, it may wish to create guidelines, regulations, or a code of practice. Together with the principles, these guidelines would form part of a comprehensive framework for the regulation of embryo testing in Australia.

---

[87] See Schwarcz SL (2009) The 'Principles' Paradox. European Business Organization Law Review 10(2): 175.

[88] Black J (2008) Forms and Paradoxes of Principles-based Regulation. Capital Markets Law Journal 3: 425, 426.

[89] Black J, Hopper M, Band C (2007) Making a Success of Principles-based Regulation. Law and Financial Markets Review 1: 191, 196.

[90] Ibid.

[91] Black J (2008) Forms and Paradoxes of Principles-based Regulation. Capital Markets Law Journal 3: 425, 449.

[92] Schwarcz SL (2009) The 'Principles' Paradox. European Business Organization Law Review 10(2): 175, 182.

8.4 Establishing a Principles-Based Regulatory Framework

This section discusses the form that a national framework for the regulation of embryo testing would take. It begins by considering whether federal legislation or uniform national legislation[93] should form the basis of this framework. I argue that the basis of a national framework for the regulation of embryo testing should lie in federal legislation. Federal legislation is preferable to uniform national legislation for two key reasons. First, the process for making uniform national legislation may lack democratic accountability, due to inadequate parliamentary scrutiny.[94] Consider, for example, the 2024 proposal by the Fertility Society of Australia and New Zealand (FSANZ) for the development of a Uniform National Fertility Law.[95] The FSANZ recommended that the Health Ministers' Meeting (HMM)[96] undertake this task, which would involve template legislation being passed by a 'lead jurisdiction', with other jurisdictions then following suit.[97] This may mean that the first opportunity a state parliament has to scrutinise a particular proposal is when it is presented with a template statute.[98] As the HMM is an executive body, it lacks the democratic accountability of parliament. This is because the deliberations of the HMM are not conducted in public and thus not transparent.[99] Further, state parliaments may feel pressured to enact a statute on which intergovernmental agreement has been reached.[100] Therefore, in the context of the FSANZ proposal, a state parliament may hesitate to reject the template statute to avoid being seen as disruptive.

The second key reason for preferring federal legislation is consistency. Federal legislation would apply equally to all states and territories. This would prevent the difficulties that may arise when gametes and embryos need to be moved across state and territory borders. As FSANZ acknowledge in their proposal for a national fertility law, state parliaments would have the capacity to modify sections of the

---

93 Uniform national legislation refers to legislation that is the same (or substantially similar) in all jurisdictions. This may be done through one jurisdiction passing a template statute and the other jurisdictions following this template.

94 Kildea P, Lynch A (2011) Entrenching 'Cooperative Federalism': Is it Time to Formalise COAG's Place in the Australian Federation? Federal Law Review 39(1): 103–130, 117.

95 Fertility Society of Australia and New Zealand (2024) Draft Findings, Recommendations and Framework for FSANZ 10 Year National Fertility Roadmap. Available via https://www.fertilitysociety.com.au/about/media-news/ Accessed 30 January 2025.

96 Note: the FSANZ proposal refers to the COAG Health Council, however this body is now known as the Health Ministers' Meeting.

97 Fertility Society of Australia and New Zealand (2024) Draft Findings, Recommendations and Framework for FSANZ 10 Year National Fertility Roadmap, 8. Available via https://www.fertilitysociety.com.au/about/media-news/ Accessed 30 January 2025.

98 Kildea P, Lynch A (2011) Entrenching 'Cooperative Federalism': Is it Time to Formalise COAG's Place in the Australian Federation? Federal Law Review 39(1): 103–130, 117.

99 Appleby G, Davis M et al. (2024) Australian Public Law. Oxford University Press, p 285.

100 Kildea P, Lynch A (2011) Entrenching 'Cooperative Federalism': Is it Time to Formalise COAG's Place in the Australian Federation? Federal Law Review 39(1): 103–130, 117.

law to accommodate their own perspectives.[101] This may result in legislation around Australia that is no more consistent than what is currently in force. Federal legislation, on the other hand, would bring with it the same benefits as uniform national legislation, such as harmonisation of the law across Australia and the opportunity for central oversight,[102] but with increased consistency and transparency.

The proposed federal legislation would contain the principles outlined earlier in this chapter.[103] Further, all legislative provisions pertaining to embryo testing would be in one statute, thus simplifying the legislative landscape.[104] Enshrining the principles in legislation means that they are subject to Parliamentary scrutiny and therefore reflect the interests of society. The risk of this approach is that the principles may end up being more conservative than envisaged in this book. The section now turns to the constitutional mechanisms for enacting federal legislation to regulate embryo testing. Two options for the establishment of a national framework will be discussed. The first concerns the use of the 'reference power' in s 51(xxxvii) of the *Australian Constitution*. The second possible method is through the use of the 'corporations power' in s 51(xx) of the *Australian Constitution*.[105] Following on from this, consideration is given to the establishment of a federal regulator for ART and the issue of enforcement.

8.4.1 The Reference Power

As I have argued that federal legislation should be the basis of a national regulatory framework for embryo testing, this section looks at one of the possible mechanisms through which such legislation may be brought about. Section 51(xxxvii) of the *Australian Constitution* provides that the Commonwealth Parliament may legislate with regard to any matters referred to it 'by the Parliament of any State or States, but so that the law shall extend only to States by whose Parliaments the matter is referred, or which afterwards adopts the law'. To create a national framework through

---

[101] Fertility Society of Australia and New Zealand (2024) Draft Findings, Recommendations and Framework for FSANZ 10 Year National Fertility Roadmap, 8. Available via https://www.fertilitysociety.com.au/about/media-news/ Accessed 30 January 2025.

[102] Ibid.

[103] It is likely that the legislation would also contain other principles to regulate other areas of ART. However, the interaction of the principles outlined in this chapter with the other possible principles in the proposed statute is beyond the scope of this book.

[104] This would mean that s 11 of the *Researching Involving Human Embryos Act 2002* (Cth) could be repealed and the 2002 Act would be solely focused on research.

[105] A third possible way of establishing a national framework is through the use of the 'executive power' in s 61 of the *Constitution*, used in conjunction with the 'incidental power' in s 51(xxxix). However, this is unlikely to be successful, as it would be difficult to argue that embryo testing, or even ART generally, is incidental to 'the execution of any power' given to the Commonwealth by the *Constitution*. As the High Court noted in *Pape v Commissioner for Taxation* (2009) 238 CLR 1 [228], s 51(xxxix) does not permit the Commonwealth Parliament to legislate on 'any subject which the Executive regards as of national interest and concern'.

the use of the reference power would require agreement from all states, otherwise the resultant laws will not have national application. The use of the reference power is the preferred method by which to achieve national consistency, as it would indicate that all states support a national approach to embryo testing specifically and perhaps ART more generally. Further, if a majority of states are in agreement about the need for a national framework for the regulation of ART as a whole, there is the option (as stated in s 51(xxxvii) of the Constitution) for any states or territories that do not wish to participate initially, to later adopt the law. Furthermore, the law would be simplified, as the six states with legislation would repeal their ART statutes.

Moens and Trone suggest that the subject of a reference to the Commonwealth Parliament, in this case embryo testing, may be 'expressed in general terms'.[106] Further, they state that when a matter is referred under s 51(xxxvii), specific limitations may be placed upon the reach of the resulting legislation.[107] Therefore, it would be possible for National Cabinet[108] to make a reference under s 51(xxxvii) pertaining to embryo testing and any other aspects of ART on which agreement could be reached, while leaving to the states any issues on which no agreement is possible.

It is questionable whether use of the reference power would be a realistic way to bring about national legislation on this matter. As the reference power was used in 2002 to request federal legislation on human cloning and embryo research, there is the possibility that COAG did not consider embryo testing to be a matter on which nationally applicable legislation was required. However, with advances in embryo testing COAG's successor, National Cabinet, may now think differently.

8.4.2 The Corporations Power

If it is not possible to implement a national regulatory framework for embryo testing through the use of the reference power, an alternative mechanism for achieving the desired result may be through the use of the corporations power in s 51(xx) of the *Constitution*. The corporations power gives the Commonwealth Parliament the authority to legislate with regard to 'foreign corporations, and trading or financial corporations formed within the limits of the Commonwealth'. Therefore, rather than requiring the agreement of the states, the Commonwealth Parliament can use the corporations power to make laws that apply across Australia. While this may be the politically simpler option, it may be the subject of legal challenges from the states if they do not agree with the provisions of the resulting legislation. A further drawback of the corporations power is that the states may choose to retain their own statutes, leading to an extra layer of regulation; however, the existing state legislation will

[106] Moens G, Trone J (2012) Lumb, Moens & Trone, The Constitution of the Commonwealth of Australia Annotated, 9th ed. LexisNexis Butterworths, 209 [397].

[107] Ibid 210 [397].

[108] On 29 May 2020, National Cabinet replaced the Council of Australian Governments (COAG).

be rendered inoperative to the extent that its provisions conflict with the proposed Commonwealth legislation.[109]

In the *WorkChoices* case,[110] it was held that s 51(xx) bestows upon the Commonwealth Parliament a broad power to legislate on any matters related to the 'business functions and activities of constitutional corporations'.[111] The High Court endorsed the view of Gaudron J in *Re Pacific Coal*, where she stated:

> I have no doubt that the power conferred by s 51(xx) of the Constitution extends to the regulation of the activities, functions, relationships and the business of a corporation described in that sub-section, the creation of rights, and privileges belonging to such a corporation, the imposition of obligations on it and, in respect of those matters, to the regulation of the conduct of those through whom it acts, its employees and shareholders and, also the regulation of those whose conduct is capable of affecting its activities ...[112]

As embryo testing is an activity that ART clinics undertake it falls within the scope of s 51(xx), provided such clinics can be deemed constitutional corporations. If they can be said to be constitutional corporations, Commonwealth legislation regulating their practices may be enacted. To fall within the scope of s 51(xx), a corporation must engage in 'substantial' trading or financial activities.[113] As ART providers are not financial corporations, the focus of this discussion will be on whether the provision of ART services qualify as substantial trading activities for the purposes of s 51(xx). As embryo testing is not covered by government funding in Australia, the provision of such testing would qualify as a trading activity, as it is a service a clinic is selling. As Mason J pointed out in the *WA Football League* case,[114] 'trading' includes all activities which produce revenue.[115] Even though less specialised ART services such as IVF attract a rebate from Medicare, clinics such as IVF Australia for example, will still fall within s 51(xx) as their activities produce revenue.[116] Therefore, ART clinics can be said to be constitutional corporations.

More complex may be whether collaborations between a public hospital and a private ART provider fall under s 51(xx) (e.g. the Fertility Unit at the Royal Prince Alfred Hospital in Sydney, which is a collaboration with Genea).[117] In *E v Australian Red Cross Society*,[118] the Royal Prince Alfred Hospital was considered to be a trading

---

[109] *Australian Constitution* s 109.

[110] *New South Wales v Commonwealth* (2006) 229 CLR 1.

[111] Ibid [177].

[112] *Re Pacific Coal Pty Ltd; Ex parte Construction, Forestry, Mining and Energy Union* (2000) 203 CLR 346 [83].

[113] Hanks P, Gordon F, Hill G (2012) Constitutional Law in Australia, 3rd ed. LexisNexis Butterworths, 401.

[114] *R v Federal Court of Australia; Ex parte Western Australian National Football League* (1979) 143 CLR 190.

[115] Ibid 235.

[116] This is further supported by the fact that IVF Australia's parent company Virtus Health is floated on the Australian Stock Exchange.

[117] See https://www.slhd.nsw.gov.au/rpa/fertility/about.html Accessed 24 January 2025.

[118] (1991) 27 FCR 310.

corporation within the meaning of s 4 of the *Trade Practices Act 1974* (Cth).[119] This is because when considering whether a corporation falls within s 51(xx), a court will take into account more than just its 'predominant and characteristic activity'.[120] Therefore, it may be argued that as the Royal Prince Alfred Hospital was projected to generate over $155 million in revenue during the 2024–2025 financial year, it would qualify as a trading corporation.[121] The same conclusion may be reached by arguing that the public hospital falls within the scope of s 51(xx) because of their partnership with the private company. This is because the hospital would be an entity 'whose conduct is or is capable of affecting [the] activities'[122] of the private provider. As a consequence, the corporations power may be used in order to pass federal legislation regulating embryo testing.

8.4.3 An ART Regulator for Australia

In a similar vein to the *Human Fertilisation and Embryology Act 1990* (UK), the proposed legislation would establish a national regulator for ART. This regulator would have three functions: oversight of clinics, an advisory function and a public education function. As noted earlier in this chapter, there is only one state with a dedicated ART regulator in Australia: the RTC in Western Australia. In addition, there exists the RTAC of the FSANZ. The expertise of the staff employed by both of these organisations could be transferred to the new regulator. It may even be possible for the RTAC to become the new ART regulator for Australia. As the RTAC already plays an important role in compliance, it would simply need to be equipped with the additional resources to enable it to successfully perform the proposed advisory and public education functions.

Rather than the RTAC Code of Practice directing clinics to the NHMRC guidelines (as is currently the case), the RTAC Code of Practice could be expanded to include guidelines that help clinics interpret the legislative principles. Should clinics have questions pertaining to either the Code of Practice or the legislation, the RTAC will have the capacity, as part of its advisory function, to assist with these queries. As part of its public education function, the RTAC would be tasked with raising its profile in the ART sector (for example with patients), promoting brand awareness with the public at large, and also educating the public about advances in embryo testing and ART generally.

---

[119] Ibid 345.

[120] *R v Federal Court of Australia; Ex parte Western Australian National Football League* (1979) 143 CLR 190, 213 quoting *R v Trade Practices Tribunal; Ex parte St George County Council* (1974) 130 CLR 533.

[121] Sydney Local Health District (2024) Royal Prince Alfred Hospital: Initial Budget Allocation Financial Year 2024–2025. https://slhd.health.nsw.gov.au/sites/default/files/2024-07/Budget-RPA-Hospital.pdf Accessed 24 January 2025.

[122] *Re Pacific Coal Pty Ltd; Ex parte Construction, Forestry, Mining and Energy Union* (2000) 203 CLR 346 [83].

One possible problem with the RTAC moving into this new role, is that it is currently a committee of the FSANZ. To undertake this new proposed regulatory role, it would therefore need a greater degree of autonomy than it is currently afforded as a committee of a larger organisation. At present, the RTAC is wholly funded by the FSANZ. Under the proposed model, it would receive federal funding and would therefore be financially independent from the FSANZ, instead operating as an arm's length body. As a result, it may need a new name to reflect its status as a regulator, rather than a committee. As this would make the current RTAC redundant, it is suggested that the committee be disbanded, although this decision is ultimately up to the FSANZ.

8.4.4 Enforcement

In the event that a clinic does not comply with the principles contained in the proposed legislation and this non-compliance is deliberate or a consequence of carelessness, enforcement mechanisms need to be available to the regulator. This section outlines a possible enforcement regime for breaches of the embryo testing principles. For other ART related offences, a different sanctions regime may be more appropriate. The current framework provides for a combination of fines and prison terms for those who do not comply with the law. For example, in Victoria, the current fine for carrying out non-medical sex selection is $47,421.60.[123] The legislation also leaves open the option of a term of imprisonment of two years in lieu of or in addition to the fine.[124] In Western Australia, the same offence attracts a possible jail term of five years.[125] As stated earlier in this chapter, one of the benefits of a national system of regulation is that penalties for non-compliance will be uniform across Australia. In addition to the above penalties, clinics that conduct unauthorised tests on embryos are liable to lose their RTAC accreditation. If a clinic continues to operate without accreditation it will commit an offence pursuant to the *Research Involving Human Embryos Act 2002* (Cth) s 11 which carries a penalty of five years imprisonment.[126]

While prison terms may seem unnecessary, they may be an effective deterrent for preventing continued breaches of the law, engaging in practices that endanger the safety of patients and those that put the health and wellbeing of the future child at risk. However, they should only be used as a last resort. Baldwin et al. note that securing a criminal prosecution is 'so costly in time and money that selective use of less formal mechanisms may produce higher levels of compliance for a given

[123] This amount is equivalent to 240 penalty units: *Assisted Reproductive Treatment Act 2008* (Vic) s 28(1).

[124] Ibid.

[125] *Human Reproductive Technology Act 1991* (WA) s 7(2).

[126] One of the benefits of commonwealth legislation covering ART as a whole is that s 11 of the *Research Involving Human Embryos Act 2002* (Cth) could be repealed. This would mean all legislative provisions pertaining to ART are in one place and thus easily accessible to patients and clinic staff.

level of state expenditure than is possible with routine prosecution'.[127] Thus rather than imprisonment being the first and only sanction available, it should be the last option in a list of increasing severity. Ayres and Braithwaite refer to this kind of enforcement strategy as an 'enforcement pyramid' because most of the actions taken by the regulator are at the base of the pyramid, with the more severe sanctions (those at the top of the pyramid) being used less often.[128] The sanctions available to regulator and the way they will be deployed would be detailed in the legislation. The Code of Practice could then reiterate this information so that it is easily accessible to all stakeholders, as well as providing further details about the regulator's enforcement strategy. As Ayres and Braithwaite note, compliance is more likely to occur when regulated entities can see a clear enforcement pyramid, and enforcement action by the regulator is transparent.[129]

A possible penalty for a first offence may be a written warning. It is important with a principles-based regime that the initial penalty is not too harsh, as this may give rise to the principles paradox detailed above. If the clinic continues to offend, a fine may be imposed. This fine should be significant in order to act as a deterrent. In order to deter further non-compliance, the penalty for a third offence should be more severe, such as suspension of accreditation for a period of time. This would be a significant sanction for two reasons; first, while the clinic's accreditation is suspended they would be unable to operate and would therefore have to forgo revenue. Secondly, having to explain to existing and prospective patients why they have to shut down for a specific period of time would have a serious impact on a clinic's reputation.[130] During the time the clinic is not operating, it would have the opportunity to provide further compliance training to staff. Should senior staff require assistance with this task, the proposed regulator would be on hand to assist as part of its advisory function. If none of these sanctions are effective at eliminating the undesired behaviour, a term of imprisonment may be an option. As Ayres and Braithwaite note, the key to 'successful regulation is to establish a synergy between punishment and persuasion.'[131]

---

[127] Baldwin R, Cave M, Lodge M (2012) Understanding Regulation: Theory, Strategy, and Practice, 2nd ed. Oxford University Press, 240.

[128] Ayres I, Braithwaite J (1992) Responsive Regulation. Oxford University Press, 35.

[129] Ibid.

[130] Ayres and Braithwaite refer to this as 'reputational deterrence'. Ibid 30. This assumes clinics would be honest with their patients about why they are not operating, which may not always be the case.

[131] Ibid 25.

8.5 Conclusion

This chapter has argued that Australia needs a national, principles-based framework for the regulation of embryo testing. While the focus of this book is the regulation of embryo testing, a number of the arguments and suggestions for reform may be applied to ART as a whole.

This chapter has identified a number of problems with the current regulatory framework that highlight the need for reform. These include a lack of consistency between the provisions that apply in each state, the different levels of support offered to clinics depending on their location and the disparate sanctions imposed for non-compliance. The chapter then turned to the advantages of taking a national approach. These include the introduction of laws that apply across Australia, so patients receive equal treatment regardless of where they are based, financial benefits stemming from harmonisation and reducing the compliance burden on clinics. Further advantages stemming from the implementation of a national framework for the regulation of embryo testing were illustrated by considering the regulatory framework for ART that exists in the UK. This discussion highlighted the importance of establishing a specialist regulator for ART in Australia that encourages transparency and provides greater support for clinics.

Having justified the necessity of a national framework for the regulation of embryo testing, the discussion turned to the reasons a proposed framework should be principles-based as opposed to rules-based. A principles-based model provides a level of flexibility that enables it to easily keep pace with innovation, without constantly being amended. Further, broad principles may also help facilitate a consultative, joint decision-making process between doctors and their patients.

It has been argued that there are three principles on which the regulation of embryo testing should be based. The first principle is that embryo testing should preserve the future child's right to an open future. The second principle is that prospective parents should be permitted to act in a procreatively beneficent manner. The second principle may be overridden by the first principle in the event that they conflict. This chapter has shown how these principles may be applied in cases of selection against disabling conditions, selection for abilities, selection for disability and non-medical sex selection. The discussion then moved on to the selection of HLA matched embryos that have the potential to save the life of a sick older child. In this context it was noted that HLA typing does not sit comfortably within the framework of the first and second principles. Therefore, a third principle, the balance principle, was introduced. It provides that where embryo testing is carried out for the purposes of creating a tissue-matched embryo, regard must be had for the welfare of the future child.

The proposed regulatory framework would take the form of federal legislation containing the three principles. The mechanism through which commonwealth legislation could be enacted is through the use of the corporations power in s 51(xx) of the *Australian Constitution*. The legislation would establish a national regulator with three functions: advisory, compliance and public education. The regulator would also be tasked with producing a code of practice that provides clinics with additional guidance on how to interpret the principles. Further, the regulator would have enforcement powers and may penalise clinics in accordance with the enforcement strategy detailed in the legislation.

Index

M. de Souza, *The Regulation of Embryo Testing in Australia*,
https://doi.org/10.1007/978-981-96-5902-9

The manufacturer's authorised representative in the EU is Springer Nature Customer Service Centre GmbH, Europaplatz 3, 69115 Heidelberg, Germany. If you have any concerns regarding our products, please contact ProductSafety@springernature.com

Printed and bound by CPI Group (UK) Ltd, Croydon, CR0 4YY
30/09/2026
02225296-0004